Friendship, Robots, and Social Media

Various emerging technologies, from social robotics to social media, appeal to our desire for social interactions, while avoiding some of the risks and costs of face-to-face human interaction. But can they offer us real friendship? In this book, Alexis Elder outlines a theory of friendship drawing on Aristotle and contemporary work on social ontology, and then uses it to evaluate the real value of social robotics and emerging social technologies.

In Part I of the book Elder develops a robust and rigorous ontology of friendship: what it is, how it functions, what harms it, and how it relates to familiar ethical and philosophical questions about character, value, and well-being. In Part II she applies this ontology to emerging trends in social robotics and human-robot interaction, including robotic companions for lonely seniors, therapeutic robots used to teach social skills to children on the autism spectrum, and companionate robots currently being developed for consumer markets. Elder articulates the moral hazards presented by these robots, while at the same time acknowledging their real and measurable benefits. In Part III she shifts her focus to connections between real people, especially those enabled by social media. Arguing against critics who have charged that these new communication technologies are weakening our social connections, Elder explores ways in which text messaging, videochats, Facebook, and Snapchat are enabling us to develop, sustain, and enrich our friendship in new and meaningful ways.

Alexis M. Elder is an assistant professor of philosophy at the University of Minnesota Duluth. Her research focuses on friendship and social technologies. Her publications include "Excellent Online Friendships: An Aristotelian Defense of Social Media" in *Ethics and Information Technology*, and "Zhuangzi on Friendship and Death" in *Southern Journal of Philosophy*.

Routledge Research in Applied Ethics

1 **Vulnerability, Autonomy and Applied Ethics**
Edited by Christine Straehle

2 **Refugees and the Ethics of Forced Displacement**
Serena Parekh

3 **Procreation, Parenthood, and Education Rights**
Ethical and Philosophical Issues
Edited by Jaime Ahlberg and Michael Chobli

4 **The Ethics of Climate Engineering**
Solar Radiation Management and Non-Ideal Justice
Toby Svoboda

5 **Corporal Punishment**
A Philosophical Assessment
Patrick Lenta

6 **Hobbesian Applied Ethics and Public Policy**
Edited by Shane D. Courtland

7 **Does the Pro-Life Worldview Make Sense?**
Abortion, Hell, and Violence Against Abortion Doctors
Stephen Kershnar

8 **The Injustice of Punishment**
Bruce N. Waller

9 **Friendship, Robots, and Social Media**
False Friends and Second Selves
Alexis M. Elder

Friendship, Robots, and Social Media

False Friends and Second Selves

Alexis M. Elder

Routledge
Taylor & Francis Group
LONDON AND NEW YORK

First published 2018 by Routledge

2 Park Square, Milton Park, Abingdon, Oxfordshire OX14 4RN
52 Vanderbilt Avenue, New York, NY 10017

Routledge is an imprint of the Taylor & Francis Group, an informa business

First issued in paperback 2019

Library of Congress Cataloging-in-Publication Data
Names: Elder, Alexis M., author.
Title: Friendship, robots, and social media : false friends and second selves / Alexis M. Elder.
Description: 1 [edition]. | New York : Routledge, 2018. | Series: Routledge research in applied ethics ; 9 | Includes bibliographical references and index.
Identifiers: LCCN 2017051253 | ISBN 9781138065666 (hardback : alk. paper)
Subjects: LCSH: Social interaction. | Friendship. | Robots. | Social media.
Classification: LCC HM1111 .E525 2018 | DDC 302—dc23
LC record available at https://lccn.loc.gov/2017051253

ISBN: 978-1-138-06566-6 (hbk)
ISBN: 978-0-367-88943-2 (pbk)

Typeset in Sabon
by Apex CoVantage, LLC

Contents

Acknowledgements

I am deeply indebted to a great many generous individuals who have supported me throughout this project, more than I could possibly list here. But I will try to name at least a few.

Richard Volkman, who has been encouraging me since my undergraduate days to explore philosophy, Aristotle, ethics of friendship, and ethics of technology. Terrell Ward Bynum for introducing me to computer ethics. Fran Grodzinsky, for giving me the idea for at least one chapter and offering sage advice and supportive feedback at many critical junctures. Heidi Lockwood, for urging me to turn a chance comment in a Q&A into a paper that then grew into a research project in social robotics. Marty Wolf for his mentorship in computer ethics. The ETHICOMP research community for many wonderful and inspiring conferences. IACAP and SPT conference-goers likewise.

The Philosophy and Women's Studies departments at Southern Connecticut State University proved to be grounds for many rich and rewarding conversations that contributed to the development of the ideas in this book: I am indebted to Xiaomei Yang, Armen Marsoobian, Eric Cavallero, Ken Gatzke, Rex Gilliland, Krystyna Gorniak, Chelsea Harry, David Pettigrew, Sarah Roe, and Yi-Chun Tricia Lin, in addition to those already mentioned. Sheila Magnotti, the SCSU Philosophy Department's amazing secretary, is not only phenomenally helpful but also my Snapchat buddy.

Many thanks to Mark Lowe and the Quinebaug Valley Community College community, who had me out to give several talks that turned into wonderful conversations with faculty, administrators, and students alike on the values and risks of social technologies. My fabulous colleagues at the University of Minnesota Duluth have helped me to develop this research in new and unexpected directions, including much helpful guidance on the literature in cognitive science and behavioral economics, as well as social media, robotics, ethics, and yet more Aristotle: Jeanine and Bob Schroer, Jason Ford, Sean Walsh, and Eve Rabinoff, not to mention two terrifically helpful administrative specialists, Bridget Park and Terry Estep. Shannon Vallor, John Sullins, and the audience at the 2016 Pacific APA contributed to a stimulating conversation about social robotics.

I am indebted to the students in my classes at the University of Connecticut, Southern Connecticut State University, Quinnipiac University, and the University of Minnesota Duluth, with whom I enjoyed too many terrific discussions to list here and who greatly enriched my thinking about these complex issues. You all are the best students a teacher could ask for. I am also thankful for my colleagues and classmates at the University of Connecticut, with whom I shared many stimulating conversations about the nature of friendship, especially Paul Bloomfield.

And last but certainly not least, Thomas Dorr, steadfast cheerleader, supplier of tamales while research and writing kept me out of the kitchen, extraordinarily conscientious unpaid research assistant and reference management support, and friend.

An earlier version of Chapter 3 was published in *Ratio* (2014) Volume 27 (1), pp. 84–99 under the title "Why Bad People Can't Be Good Friends". It appears here with permission of John Wiley and Sons.

Portions of Chapter 8 were published in *Ethics and Information Technology* (2014) Volume 16 (4), pp. 287–97 under the title "Excellent Online Friendships: An Aristotelian Defense of Social Media". It appears here with permission of Springer.

Introduction

1. Human Nature, Human Connections, Human Creations

Human beings are remarkable for both our social and technical inclinations. These two interests come together in the realm of social technologies, those tools we invent and use to address our social needs. We are inventive and clever creatures, and these features are very much in evidence here. We have created a staggering variety of technologies, from writing systems and telephones to Facebook, Twitter, and Snapchat, to a rapidly-expanding array of social robotics. But cleverness and wisdom do not always coincide. There is little doubt that the majority of social technologies are intended to benefit users. However, when it comes to the social realm, our needs are complicated. Not every well-meant invention, even when it works as intended, ends up being good for us in the long run. At the same time, to reject our innovative tendencies in the social realm seems foolish. It can be tempting to criticize new technologies by hearkening back to some Golden Age of pre-technological harmony. But this seems both inaccurate in terms of actual history and dismissive of our nature as tool-users and tool innovators. Rather than shy away from our cleverness, we should use it to our advantage. But we should take care to integrate it with a clearer understanding of the subject matter with which it deals.

In this book, I take the view that social technologies can contribute to good lives, but that we should reflect on what these good lives consist in, so as to successfully practice ethical design, deployment, and use of these technologies. In order to make the case for this, I lay out, in this Introduction, some key terms and concepts. As is so often the case in philosophy, the trickiest part will not be introducing novel concepts, but in clarifying everyday or near-everyday ones, in order to make headway on topics that many find confusing or controversial.

2. Clarifying Concepts

To start with, take the question of what makes a life good one. One way to answer the question is by way of a tradition with roots in ancient Greece. The ancient Greek philosophers, and modern virtue ethicists inspired by them, use the term *eudaimonia* to refer to something like this. Aristotle, from whose work I take much inspiration, developed a rich and complex account of what it takes for human beings to live well. The Greek term he used for this desirable phenomenon, *eudaimonia*, is translated into English, variously, as: the good life, living well, well-being, being blessed, happiness, and flourishing, and within the framework of Aristotelian virtue ethics it has overtones of each. Contemporary virtue ethicists—including myself—who take the ancient Greeks to have been on to something important find it helpful to link these concepts together. We also follow Aristotle in thinking that *eudaimonia* is conceptually linked with virtue. Aristotle offers an account of *eudaimonia* in which a creature's ultimate good consists of living well as the kind of thing it is. This explains why terms like flourishing, living well, and well-being are often used to translate *eudaimonia*. "Happiness" gets added to the mix, not by referring to hedonist pleasures, but as an alternate way of getting at the kind of enjoyment found in performing one's work wholeheartedly and well (even if this is not always pleasant). Blessedness, a translation occasionally favored by Terence Irwin (Aristotle 1999), matters here because *eudaimonia* requires elements of good fortune not always under the individual's control: one cannot live well without the right external goods, such as access to adequate nutrition, supportive political structure and community, and enough wealth and resources to take care of basic necessities. Virtues, in an Aristotelian account, do not imply prudishness or purity, even though these are things we often associate with the word "virtue" today. Instead, they are the characteristics necessary for a creature to function well as the kind of thing it is. A knife, for instance, needs a sharp edge in order to cut well, and so sharpness is a virtue of knives. By linking all of these ideas together, we end up with a framework for making sense of a good life as something that is deeply connected to our nature, so that in order to understand what it would be for us to live well, we need to understand what sort of creatures we are.

Fully-developed accounts of what our nature is and what the virtues are that will help us to live according to this nature are abundant in academic philosophy, and it is not my intention here to arbitrate between them or to make my account dependent on the details of one particular account over others. Instead, I aim to make headway by three means: two substantive, one methodological. The methodological element is a commitment to reflecting on what constitutes a good life when we encounter difficult ethical questions in some particular context where various goods or ethical goals appear to be in conflict. This is used throughout the book to shed light on various social and individual goods and technologies.

The substantive elements involve the importance of sociality and technology to human beings. On the importance of the social, Aristotle noted in the *Politics* (Book 1, Chapter 2, Aristotle 2017, 4, 1253a) and elsewhere that human beings are inherently social animals. Like bees or wolves, we live in organized groups and characteristically work together to accomplish tasks larger and more complex than those manageable by a solitary individual. Thus, even if any given person were not terribly inclined to enjoy others' company, there might be a case to be made for social connection as constitutive of functioning well. But as it happens, I think that for most of us, social connection is also inherently valuable, a point which can be brought out by reflecting on a thought experiment inspired by Aristotle. He claimed, "It would be "absurd to make the blessed [*eudaimon*] person solitary. For no one would choose to have all [other] goods and yet be alone, since a human being is a political [animal], tending by nature to live together with others" (Aristotle 1999, 148, 1169b, 15–20). Note that this involves not only a biological claim about human social organizations, but also an appeal to individual choice rooted in what one values.

Imagine you were given a choice between a life in which you enjoyed all of the material goods, but no human connections, and one in which you had access to all of these, plus friends, family, and community. To most of us, this is not a difficult decision: we opt for the life with social connection over that without. "But wait", one might object. "That's just because the second option involves more than the first". But the second option only seems more desirable because it has more in it that we find valuable: an option that included lots of an element a person considered worthless would not thereby be considered more choice worthy. And suppose we modify the options slightly. You can choose between a life with all the material goods you could ever want but no human connection, and one in which your basic needs are satisfied but you do not always get *every*thing you want. But in this second scenario, you still have a rich social life. If the choice between this one and the socially barren one still seems obvious to you, then not only does it seem that the latter choice includes goods missing from the former, but that these goods are so valuable they are worth trading for greater quantities of other goods. This suggests that human connection is not just *a* good, but an irreplaceable element of the best human lives (or at least, the best lives for most humans). Thus, we have both objective and subjective reason to think that social lives matter to us, even if they would not matter so much for or to crocodiles or wolverines. This does not mean that *every* connection is good. But it does imply that good connections are good because they function well *as* social connections, not because they are efficient conveyers of value that could in principle be obtained elsewhere. What we need, then, is an account of what makes such connections what they are, so that we can take what steps we can to help them to function well. That is the topic of Part I of this book.

The second substantive element that I rely on here has to do with the importance of technology to good human lives. Here, philosophers and other thinkers have varied widely, from Luddite rejection of modern technology to full-throated technological optimism, the sort that thinks any social ill can be overcome by a clever enough invention. Some, like Albert Borgmann and advocates of the "slow tech" approach, endorse some kinds of technology (especially those in which human beings are integrated throughout the process of using these artifacts) but not others (especially those exemplifying what (Borgmann 1984) terms the 'device paradigm', where the artifact "just works" for the user while its functioning remains opaque). But like any good Aristotelian, I think the right amount lies somewhere in the middle and is not reducible to a formula or a unified explanation, as on Borgmann's account. Furthermore, what the Luddite misses is that innovative tool use, even when socially transformative, is not "unnatural" for human beings. Judging by our history, it is in fact entirely normal for human societies to vary widely and change dramatically over time as technological practices and artifacts are introduced, modified, and adapted. This does not, of course, always make such changes *good*. But it does mean that if we are to take human nature seriously in our considerations about what is ethical, we need to embrace rather than reject this aspect of our humanity. What makes any given technological innovation good or bad has to be specified by something else. The natural thing for the Aristotle-inspired ethicist to say is that innovation is an excellence when it coheres with the other excellences: when a given technology helps us to excel at being human, all things considered. So, in order to deliver a verdict on any given technology, we have to examine how well it integrates with our other values. For social technologies, this means thinking about what excellence in the social realm looks like.

3. Theorizing About Friendship

I tackle this by first laying out a detailed account of a notoriously amorphous concept: friendship. My aim here is not to convince the reader *that* friendship is an important human good. I take that as a premise. (This does not mean it needs to be central to every human life, any more than thinking that art is an important human good means thinking every person needs to devote their life to art.) Rather, it is to reflect clearly and carefully on a number of apparently puzzling features of friendship, in order to better understand a complicated social phenomenon. Furthermore, as this is ultimately being done in the service of providing ethical guidance on social technologies, I focus on articulating the nature of our *ideal* of friendship. While empirical data about actual human practices can be helpful, we need a theory of what is valuable in order to get from a pile of facts to a prescription for what to do.

Why friendship? In the field of philosophy, friendship has been given relatively little attention in recent years, especially relative to other social phenomena such as governments, economies, and even families. So this might seem to be an odd place for a philosopher to look for information about valuable social connections. There is sometimes a tendency to think of friendship as relatively trivial, the kind of relationship we sometimes choose to engage in for our own amusement, after our more "serious" needs are met. But historically, friendship has often been given pride of place in the social world. From the classical Chinese philosopher Kongzi (Confucius 2003), who praised the importance of friendship to the cultivated individual, to the extensive discussion of friendship in the Buddhist text *Dhammapada* (Buddha 2000), to the chapters on friendship in Aristotle's *Nicomachean Ethics* (Aristotle 1999), on which this volume draws heavily for inspiration, thinkers from a wide range of cultures and eras have identified friendship as an important element of a good life.

In addition, friendship itself seems to be a kind of ideal of human relationship, and while even figuring out who counts as a "real" friend can be difficult, this can be explained in part because calling someone a friend seems to be, among other things, a way of saying something positive about the person and the relationship. To describe one's co-workers as friends, or one's family members, or neighbors, is to say that there is something especially valuable about these relationships, over and above their formal membership in social categories such as colleague, sibling, or fellow resident of a community. Therefore, to say something about what a friend is, tells us something about what we aim for in many of our social lives more broadly.

Lastly, many of the most striking technological innovations of our present era, from social media and smartphones to social robotics, are explicitly designed to work with our everyday desires for companionship, comraderie, and friendly interaction. Social entities like governments, corporations, and financial institutions may draw the attention because of their ability to affect our lives in highly visible ways. But the smaller, more frequent, everyday exchanges of friends may add up to a bigger and more insidious impact on our lives overall. And research suggests that social connections, especially close ones like friends and (well-functioning) family and romantic partnerships, are important predictors of long-term well-being (Piore 2015).

Getting clear on what friendship is, then, can help us to make good decisions about the design and use of technologies that implicate it. In particular, my strategy in Part I of this volume is to figure out 1) what it means to reason well about friendship, 2) how we should think about friendships (that is, how we should understand what they are), and 3) how concerns about friendship fit in with concerns about morality, overall. In each case I start with a puzzling aspect of our ordinary beliefs and values about friendship, and attempt to organize and explain these beliefs so as to clear up the confusion without sacrificing important intuitions or beliefs about what friendship is or how it ought to go. Some of the work done here is ground-clearing: that is, by clearing up what we are talking about when we talk about friendship as an ideal at which we aim, we can make better use of available technologies, and better innovate when constructing new ones. We avoid getting tripped up on confusions about the target concept. Other aspects of the work lay foundations for specific conclusions about social media and social robotics, allowing for further development in later chapters that, while less abstract and theoretical, engage more explicitly with the technologies in question.

Because evaluating social technologies requires that we reason appropriately about our social connections, I begin Chapter 1 with a puzzle about reasoning in friendship. Reasons seem by their very nature to be repeatable and consistent in order to be rational, and yet reasons for friendships, especially reasons for making friends, maintaining friendships, or ending friendships, do not seem to be subject to expectations of consistency. At the same time, much of friendship does seem to involve the use of reasons, even in these very decisions. Some theorists attempt to explain away the lack of consistency and make friendship ultimately rational, while others try to explain away the appearance of rationality and make friendship fundamentally resistant to reasons. In this chapter, I argue that both pursue a mistaken strategy by taking friends' reasoning to focus tightly on individual friends and actions. What we find in friendship is not reasoning about individuals and how they treat us, but rather reasoning about the constitution and well-being of a complex social entity jointly composed by the friends. By reframing the issue in this way, much of the apparent puzzlement drops away, and various kinds of reasoning typical of friendship that seemed to be in tension turn out merely to be involved in different but compatible aspects of supporting the well-being of the friendship.

Having introduced this idea of a friendship as a social entity, Chapter 2 is devoted to unpacking an ontology of friendship. In addition to the reasons given in Chapter 1, introducing these entities can explain a worry about identity in friendship. There seems to be tension between thinking that close friends are, as Aristotle puts it, "other selves", and thinking that friendship can enrich us by difference as well as sameness. But theoretical resources from metaphysical accounts of the relationship between parts and the wholes they compose can be used to make sense of the idea that friends are other selves and yet different. They do so by making friends out to be different parts of a composite entity, a friendship, and in virtue of being parts be legitimately considered to be identical with the whole they compose. What emerges is an account that prioritizes interactions between and interdependence of friends over similarities between friends. Just as a heart and liver of an organism can look very different and perform very different activities while jointly composing the same organism, so friends can have different roles to play in the friendship and different strengths to contribute, so long as they each help the other to function well and are appropriately interconnected.

Lastly, I discuss the relationship between moral concerns and concerns of friendship. This is a contested issue. On the one hand, loyalty and concern for friends seems to give one reason to behave badly under certain circumstances. As the old adage goes, a friend helps you move, but a good friend helps you move a body. On the other hand, friendship seems to be benefited by the good character of friends. I reconcile this tension and argue that our ideals of friendship give us reasons internal to friendship to be concerned with our own and our friends' character by noting the importance of concern for friends' well-being to ideals of friendship. Once we connect well-being to virtuous character, as we have already seen that virtue ethics does, a concern for morality emerges out of concerns of friendship.

This portion is relatively technical, and not every concern addressed here may be of interest to every reader. Nevertheless, if you find yourself asking things like, "Why don't we just figure out what the qualities of a good friend are, so we can duplicate them in social robots?" or "Why Instagram your lunch? Why think that it should matter to a person what their friends are up to when they aren't together or at least involved in something of shared interest?" or "What does friendship have to do with morality?" then the conclusions argued for in these chapters will be of interest to you.

4. Theorizing About Technologies

On the issue of morality and moral theory, my approach is broadly eudaimonist, and specifically inspired by Aristotle's account of virtue ethics. I do not consider myself committed to everything Aristotle had to say on the subject. However, the idea of beginning ethical thinking by reflecting on what living well consists in seems to me a fine starting point, and particularly helpful when thinking about ethical issues involving new technologies. In order to decide what constitutes an ethical or unethical technology or use of same, it is often a good idea to start by thinking about what kinds of lives it enables or makes difficult, and how valuable these lives are, relative to the alternatives available. Furthermore, an Aristotelian approach has us answer this question by thinking about what kind of creature we are, and what it would mean for us to function well as that kind of thing, and these issues are quite relevant in the area of social technologies.

An approach informed by virtue ethics in the Aristotelian tradition can begin by observing that we are both social creatures, and innovators of technologies, and have been throughout history and across a wide range of cultures. The question then becomes, how can we innovate well? How can we use technology to build healthy rather than unhealthy social connections? How can we engage our reasoning and emotional capacities wisely and richly when creating and using social technologies?

In Parts II and III, with a working understanding of the nature of friendship in hand, I go on to consider this question in two rapidly developing contexts: social robotics and social media. Each of these contexts involves a particular kind of issue posed by technologies that engage our social responses. And each presents a rich array of ethical problems and emerging empirical data for use in theorizing about said problems. Although these two areas do not exhaust the array of social technologies, by highlighting two important kinds of issues that can arise in social technologies, and providing resources for engaging with these issues via a context-sensitive, empirically informed virtue ethics, they allow me to demonstrate how we can exercise practical wisdom to make good decisions about social technologies.

4.1. Social Robotics

Part II focuses on robots, in particular so-called social robotics. This broad category encompasses a variety of robots designed to engage social responses, for a wide range of ends. It can include assistive robots that help senior citizens around the house, therapeutic robots that teach social skills to children with autism spectrum disorders, entertainment robots intended for the consumer market, companionate robots that provide social experiences to residents of geriatric care facilities, and most notoriously, sex robots. While delivering a detailed analysis of all of these applications would take at least a full book, here I look at several applications, following the thread of friendship begun in Part I. Robots that engage our social responses can appear to users as friends, and based on the details of cognitive ability and circumstance, different users can be more or less aware of the fact that these machines differ significantly from many other things that present a friendly appearance.

The appearance of friendship is, in fact, sometimes explicitly sought-after by some social roboticists. But in order to assess the relationship between appearance and reality here, we need an account of what friendship *really* is. Here, my initial results about the importance of interdependence of valuing creatures with rich lives quickly lead to the conclusion that current and near-future robots are not capable of real friendship. This leads me to probe its limits with a series of thought experiments about the importance of reality and appearances in friendship. In Chapter 4, I intervene on a debate over the ethical importance of this distinction by means of a thought experiment about the relative value of friendly appearances without friendship versus with. Intuitions from this case are connected to a claim by Aristotle that false friends are analogous to false coinage. But as I go on to explore various therapeutic and entertainment uses of social robotics in subsequent chapters, the analogy yields richer and more varied results than it might first appear, and the appearance of friendship turns out to be important to social animals in many ways, even as keeping track of the distinction remains important.

I end up concluding that while social robots present the appearance rather than the reality of friendship, this does not make them automatically immoral nor irrelevant to friendship. Instead, by investigating their use in three contexts (geriatric care, autism spectrum therapies, and consumer markets), I engage in sustained exploration of the varying roles that appearances of friendship can play in helping us to live well as social creatures, from exercising and developing social capacities to reinforcing unachievable expectations to deceiving the vulnerable to activating physically beneficial physiological responses. Just as toy money can help us to develop and exercise our financial skills in a variety of games, toy people, properly deployed, can help us exercise our social muscles. But just as care needs to be taken to avoid misleading people as to the value of fake currency, the same is true of friendship.

So, given the ontology of friendship developed in Part I, I head off a series of bad rationalizations for social robotic technologies, while using an ethical framework that focuses on the importance of cultivating and exercising social abilities to articulate beneficial uses of social robots. And given the importance of friendship to our ideals of good lives, I conclude that this gives us reason to clearly signal when something is merely an appearance of friendship, rather than a real instance of the phenomenon. Friendship requires connection between people, and with the current generation of robots there is nothing with which to connect.

4.2. *Social Media*

When it comes to social media, by contrast, connection between people is the very basis of these technologies. Here, the question is not about that with which one connects, but the quality of the connection. In Part III, I engage in a systematic examination of the ethical significance of these new modes of technologically mediated connection between people.

Technologically mediated connection between people is nothing new. Depending on how broadly one construes technology, this may begin with spoken and signed language itself, and almost certainly encompasses writing, telegraphs, and telephones. The rise of computers and the Internet has enabled email, videochat, and various social media platforms, and the introduction of smartphones has expanded the array from SMS messaging to Instagram, Snapchat, WhatsApp, and more.

Beginning with Socrates's complaints about the inadequacy of the written word in the *Phaedrus*, technological skeptics have worried about the impact of communication technologies on individuals and relationships. In particular, many working in the Aristotelian tradition have expressed concerns about mediated interactions, especially computer-mediated communication, by appealing to Aristotle's claim that the highest form of friendship involves people sharing lives. If people cannot live together, then what we are left with is mere communication, not rich relationship. I begin the third and final section of this book by taking on this objection.

In Chapter 8, I note that Aristotle's own account of living together is "sharing of conversation and thought, not sharing the same pasture" like cattle. Unless we want to take on the implausible conclusion that good friends must cohabitate, we should take seriously the idea that friends can enjoy friendship without sharing every aspect of their lives. The question then becomes, what sorts of aspects *do* need to be shared for friendship, and which are possible via various forms of technological mediation?

Without going so far as to fully endorse Marshall McLuhan's famous dictum that "the medium is the message" (McLuhan 1964), it is uncontroversial that different communication channels can shape the messages communicated, and affect the quality of communications and relationships sustained by these communications thereby. So in this third section I explore different ways that various communication channels shape interactions between intimates. More sophisticated objections to technologically mediated communications often involve the relative impoverishment of these communication channels. One way they can be impoverished relative to in-person interaction is by reducing the available content. Consider, for example, the difference between written and spoken language: speech includes elements such as volume, intonation, and pace that are absent from written language. This is a challenge for message recipients, who are given less to work with than in face-to-face exchanges. Another way interaction can be impoverished is by reducing the impact of another's presence during communications, an impact to the speaker. Both of these phenomena seem to have the effect of introducing greater emotional distance between speakers, which in turn seems antithetical to the rich shared lives we associate with excellent friendship even when we reject a literal understanding of "shared living". Chapters 9 and 10 address these concerns specifically.

In Chapter 9, I evaluate asynchronous communication technologies. The ability to pick up and put down conversations at will—or to invisibly disengage from them or refuse to engage at all—seem most useful at establishing and enforcing boundaries between people, rather than enabling connection. But this assumes what turns out to be a false dichotomy between boundaries and the bonds of friendship. I conclude that a clearer account of what constitutes a good friendship has room for healthy boundaries between individuals, and that asynchronous communication technologies have an important role to play in helping people stay connected while enriching each other's lives via the pursuit of often very different activities. Thus, the ontology of friendship developed in Chapter 2 and the importance of distinguishing unification-as-blending from an organism account of shared identity via the collaboration of different parts of an interconnected whole yields substantive results about the ethical significance of communication technologies.

In Chapter 10, I tackle the issue of emotional engagement via technologically mediated communication. I argue that emoji and other non-verbal elements of computer-mediated communication are gaining in popularity because they enrich the emotional impact of messages, an unsurprising claim. But less obviously, I conclude that because they allow for emotional responsiveness within the context of boundary-preserving asynchronous communication technologies, as was argued in Chapter 9, they open up new opportunities for emotional interdependence while respecting boundaries and supporting complementary differences. In addition, many of the graphics associated with communicating emotions have a playful, irreverent appearance that, along with distance between friends, can actually support *more* emotional intimacy by making it easier to accept, admit to, and discuss difficult emotions. At the same time, these technologies have the potential to enhance self-awareness via making explicit the choice of emotional signifiers, because these elements of communication are intentional rather than involuntary, as in facial expressions during in-person interactions.

The preceding chapters have established the positive potential of social media. But this positivity is clear-minded, not the wishful thinking associated with naive technological optimism. While I draw on empirical data to support my conclusions, my argument is not that positive implications can be found in every technologically mediated interaction. Rather, it is that relationships have the potential to be enriched by these technologies, given what we already have reason to believe about human beings and the friendships they enjoy. In drawing out the positive potential of social technologies currently enjoying rapid uptake, my aim is not to give a thumbs-up to each and every use. It is to articulate what we aspire to when we embrace these technologies. Like friendship itself, by clarifying the ideal, we may stand a better chance of hitting the target at which we aim, or at least getting closer to it. At the same time, these technologies can present problems for individuals and the relationships they value when used unwisely. So, the question becomes, how can we best navigate these technologies so as to enjoy the benefits while avoiding the risks, insofar as that is possible?

In Chapter 11, the final chapter of the volume, I take a step back from my previous tight focus on the nature of friendship and turn to more foundational features of virtue ethics and individual character. I argue that virtue ethics is the most promising prospect on offer as a source of appropriate guidance for individuals when it comes to ethical use of emerging communication technologies to support one's friendships. I begin with an observation from the anthropologist Stefana Broadbent (Broadbent 2012). She argues that there now exists a new kind of moral choice faced by users of information communication technologies. Unlike most communication channels throughout history, today many of our communication channels funnel through the same devices, especially our smartphones and laptops. Because each communication technology no longer requires specialized equipment and infrastructure, almost anyone with a smartphone or laptop has, at least in principle, access to many of the same communication channels at any given time. Thus, the choice of which channel to use to communicate at any given time is no longer directly explicable in terms of simple pragmatics such as who has access to a telephone, or how long it takes for a letter to travel through the post. Rather, the choice becomes a moral one, and social norms are correspondingly emerging to offer guidance on this choice. For example, it is becoming widely accepted that it is immoral, all else being equal, to break up with a romantic partner via text message.

Although this suggests a proto-deontological approach, in which rules or principles offer uniform guidance across situations, I argue that ultimately, the details of a context make too much of a difference for these rules to be reliable, while the unpredictability of the effects (especially long-term effects) of emerging communication technologies make consequentialism an unreliable option, while virtue ethics is well-suited to take account of the kinds of variables in play and also highly relevant (as I argued in Chapter 3) to nurturing friendships. I then offer a detailed analysis of the ways that considerations about various virtues can inform choices about different sorts of communication channels.

5. Strategy for Navigating Ethical Issues

Throughout the volume, investigations into emerging social technologies uncover both risks and rewards, which become more clearly visible with a robust account of an often-vague concept—friendship—in hand. This account, in conjunction with the resources of virtue ethics broadly construed, as well as a commitment to reflecting on the value of social relations when technologies present ethical problems, can help us to successfully avoid the risks and maximize the rewards. This allows us to live good lives on our own terms, and to thrive as the creatures we are.

Bibliography

Aristotle. 1999. *Nicomachean Ethics*. 2nd ed. Translated by Terence Irwin. Indianapolis, IN: Hackett Publishing Company. https://books.google.com/books?id=-Mf5XV8q6CgC&dq.

———. 2017. *Politics: A New Translation*. Translated by C.D.C. Reeve. Indianapolis, IN: Hackett Publishing Company. https://books.google.com/books?id=WCQgDgAAQBAJ&dq.

Borgmann, Albert. 1984. *Technology and the Character of Contemporary Life: A Philosophical Inquiry*. 1st ed. Chicago: University of Chicago Press. https://books.google.com/books?id=XcxUPwAACAAJ.

Broadbent, Stefana. 2012. "Approaches to Personal Communication." In *Digital Anthropology*, edited by Heather Horst and Daniel Miller, 127–45. Berg. https://books.google.com/books?id=FVcDAAAAQBAJ.

Buddha. 2000. *Wisdom of the Buddha: The Unabridged Dhammapada*. Edited by Susan L. Rattiner. Translated by Friedrich Max Müller. Dover Thrift Editions. Mineola, NY: Dover Publications. https://books.google.com/books?id=tuibMIJPz_kC.

Confucius. 2003. *Analects: With Selections from Traditional Commentaries*. Translated by Edward Gilman Slingerland. Indianapolis, IN: Hackett Publishing Company. https://books.google.com/books?id=6DseYHSfaagC.

McLuhan, Marshall. 1964. *Understanding Media: The Extensions of Man*. Canada: McGraw-Hill. https://books.google.com/books?id=l2pRMszKikAC.

Piore, Adam. 2015. "What Technology Can't Change About Happiness: As Pills and Gadgets Proliferate, What Matters Is Still Social Connection." *Nautilus*, Accessed September 17. http://nautil.us/issue/28/2050/what-technology-cant-change-about-happiness.

Part I

Friendship

1 Repeatable Reasons, Irreplaceable Friends

1. Introduction

Can a robot be a friend? What could treating one as a friend do to your ability to be a good friend to others? To what extent is social media use harmful to our friendships? When and how can it be used to help us nurture these social connections? Questions like these are becoming increasingly relevant to us as these and other emerging social technologies are adopted throughout the world. I think that answering these questions well must involve reflecting on what it means to be a friend, what we value when we value friendship, and what conditions must obtain in order for friendships to thrive. We can make progress by clarifying commonplace but under-examined ideas, where conflicting assumptions sometimes lurk undetected. In order to reason well about social technologies, it is important to reason well about our social lives and values. That means we need an account of what this kind of reasoning looks like. In this first chapter, therefore, I develop an account of good reasoning in friendship, starting with an examination of some tensions inherent in commonplace reflections on how we think about our friends, tensions that may look familiar to the reader even if they have never been consciously articulated.

2. Conflicting Intuitions About Friendly Value

Consider three cases of reasoning in friendship.

Reasonable Rita

Reasonable Rita explains that she became friends with Fran because of a shared interest in hiking, and maintains this friendship because of Fran's kindness and insightfulness. Rita ended a friendship with Jeremy when she discovered how self-absorbed and cruel he really was, and—more reluctantly—ended a friendship with Holly after Holly developed a serious drug problem for which she has repeatedly refused help.

Particular Patrick

Particular Patrick explains that he and his friend Fred enjoy teasing each other about their special quirks, and that the two revel in an elaborate series of in-jokes. Patrick says (truthfully) that no matter how many friends he makes, none could ever be quite the same as, or ever replace, his friendship with Fred.

Partial Padma

Partial Padma is deeply loyal to her friends. She and Frida have been friends since kindergarten, and although both have changed significantly in the 30 years since, they continue to prioritize their friendship, and value their shared history. Padma reports that she has come to see many things differently, and value things she wouldn't have otherwise, because of her friendship with Frida.

Broadly, it seems important to value friends as unique and nonrepeatable individuals, and also to value their repeatable properties, such as character traits like kindness, bravery, and honesty; traits which can be displayed by more than one person, and which we hope many people will exemplify. Furthermore, friendship seems to involve both irreducible partiality, and to be best when it is objectively justifiable (and problematic when it is not). Someone's being a bad person, or a bad friend, seems a good reason to end a friendship. A theory of friendship saying that loyalty or partiality should trump considerations about repeatables in such cases seems to miss something about friendship.

Friends, that is, seem to appeal to three different kinds of reasons:

- Generic repeatables
- Particulars
- Partiality

But the three seem to be in tension with each other. We appeal to (general) reasons to justify choice of friends, whether the choice in question is the initial befriending, opting to sustain a friendship, or opting to end a friendship. Good friends seem (non-arbitrarily) rare, which could be explained by their possessing rare but valuable traits, and there is sometimes thought to be an important overlap between good friends and good people, which would imply that the objectively good qualities characteristic of such people are important in friendship. At the same time, particularity seems important. Friends aren't (or shouldn't be thought of as) fungible, but rather should be valued as unique individuals. We value some unique and nonrepeatable properties of friends. (Quirks, shared history/experience, in-jokes.) And the loss of a particular friend seems to be both a significant and irreplaceable loss . . . even when it is, on balance, the best available option, and even when you make new friends who are better for you. Meanwhile, at least some partiality seems necessary for friendship. Loyalty is an ideal of friendship. Friends' interests and values seem as though they can and should adapt to each other over time (the reasons we became friends need not be the reason we remain friends 20 years later). Friendship takes time to develop—it is not and could not be instantaneous. Shared history can be a reason to continue or to act out of friendship. And it seems we should think well of our friends, and give them the benefit of the doubt when possible.

But universality conflicts with particularity and partiality. Particularity is resistant to repeatable reasons, which can explain choice of friends and reasons to end friendship. And partiality conflicts with norms of reason, and with responsiveness to friends' intrinsic qualities.

In this chapter, I sketch a way that qualities and loyalty through change can be reconciled in friendship. I do this by taking the object of valuing characteristic of friends to be the friendship, an entity encompassing both friends. I argue that friendships thrive or wither based on the well-being of their constituent parts, and that such well-being connects friendship to the explanatory resources of eudaimonist virtue theory. This approach captures the relevant intuitions about friendship while explaining that apparent tensions between these intuitions result from misapplications of the standards relevant to different aspects of the friendship.

2.2. *Preliminary Qualifiers*

A few introductory notes about terminology and methodology are in order.

My primary aim is to make sense of reasons in friendship, and for this purpose I will need to investigate what could count as good reasons for friends to use. Thus, throughout this chapter, I will focus on ideals of friendship. To this end, something like Aristotelian character friendship suits my needs. Aristotle argues that complete friendships of virtue (dubbed "character friendship" in Cooper 1980 to emphasize what is valued in such friendships) are rare, but that such people are "perfect" friends and that other kinds of friendship earn their title because of their resemblance to the ideal (Aristotle 1999, 123, 1157a).

By "friendship value" I mean the value intuitively characteristic of friends and friendship (not, for example, identical with monetary value, or hedonic value, although a particular friendship may also involve these sorts of value). By distinguishing between cases intuitively representative of it, and intuitively at odds with it, I will develop a clearer account as the chapter proceeds.

My aim is to explain what counts as good reasons for friends, and what kinds of reasoning are consistent with ideals of friendship. Note that this is not the same thing as explaining *why* they value friendship. (Analogously, consider the distinction between explaining what goes on when people perceive the color red, and explaining why people see red as red rather than some other hue.)

I will distinguish, in what follows, between friends, which I define roughly here as "people in a friendship", and friendships, composite objects that take friends as parts and involve persistent patterns of interaction amongst these parts. Again, I start with a somewhat vague but intuitive picture of such entities, and then get clearer on what they are and what they entail by considering various theories and their implications.

Throughout the chapter, I will distinguish between what I call "repeatable" and "nonrepeatable" properties. I have in mind something like the distinction between tropes and universals in theories of properties: repeatables may be instantiated at many different times and places throughout the world, while nonrepeatables only appear—and could only appear—once. Both appear important to friendship, but seem to play different roles in friendship value.

Because my approach involves the fairly radical claim that friendships are objects with people as parts, I spend some time making this plausible, first by taking a look at the phenomena to be explained, and then by surveying what seem to be some more straightforward ways to explain them, showing why despite some initial promise, each ends up being unsatisfactory. I then show how we can explain intuitions about friendship without the pitfalls of previous approaches, and demonstrate a theoretical application in a particularly thorny area.

A survey of some theories of friendship will show how these varieties of reasons can be used to motivate several approaches to characterizing friendship value. Before embarking on this, however, a brief overview of a major historical theory of friendship on which many contemporary accounts draw will be helpful. Note, however, that even with this shared historical basis, extant modern theories disagree about how best to interpret the account, with various Aristotelians supporting universality, particularity, and even (to some degree) partiality.

Aristotle divides friendships into friendships of utility, pleasure, and virtue. In virtue friendship, the most complete form, friends are valued in and of themselves, just as virtue is (on Aristotle's account) good in itself. Pleasure and utility friends are valued for properties that are pleasant or useful to the valuer (*Nicomachean Ethics*, Books VIII and IX). Aristotle thus draws a distinction between intrinsic and instrumental value, and between valuing a property and a person, and connects the intrinsic value characteristic of friendship with valuing a person, instrumental value with valuing a property. However, he thinks virtuous people are the only ones eligible for virtue friendships, thus making repeatable properties apparently central to his theory, a move that may not sit well with the above distinctions. Subsequent theorists (Whiting 1991; Badhwar 1991; 2005;

Nussbaum 1997; Levine 1999) have tended to prioritize either intrinsic value of the person, or repeatable value of character traits, but not both at once. Or, where both get priority, they are taken to be in tension.

Repeatable and nonrepeatable properties, and objective versus partial valuation, then, will be the focus of the next three sections, with arguments in favor of the primacy of each considered in turn. Each of these approaches, I will argue, seems unsatisfactory.

3. Considerations in Favor of Valuing Repeatable Properties

The thought that we value our friends because of their good qualities seems, on one level, consistent with Aristotle's claim that the best friends are people we find "good in themselves", and explains why we think character is important in choosing friends. This is the kind of reasoning we saw in the case of Reasonable Rita. Thus, some theories prioritize the repeatable over the nonrepeatable as objects of friendship value.

In the *Symposium*, Socrates describes the process of loving a person as gradually learning to abstract away from people's particulars to better appreciate the goodness inherent in them. Although we start out being attracted to individuals, he says that a reasonable person "should realize next that the beauty of any one body hardly differs from that of any other body . . . it's very foolish of him not to regard the beauty of all bodies as absolutely identical" (Plato 2009, 53, 210b). Individual people are useful as rungs on a ladder, in his memorable analogy, to help a person appreciate the good properties they instantiate.

This may sound somewhat cold and unfriendly, but a more sophisticated version of this approach is adopted in Jennifer Whiting's "generic" account of friendship valuing (Whiting 1991). In "Impersonal Friends", Whiting presents a dilemma about reasons in friendship: either valuing of friends is "brute" and inexplicable, or else it is based on reasons. If it is based on reasons then it seems the reasons must be somehow generic, repeatable, and consistently applicable. We may also value distinctive features of a particular friend, but for instrumental reasons that make it easier or more likely that the two of you spend time together. This can include features like shared tastes, as well as random happenstance—that I happened to meet this person first, or lack time to develop another friendship at the moment.

She bites the bullet in making friendship impersonal, as she grants in her title. But, she argues, the explanatory advantages of this approach are great. She charges that major competitors are unsatisfactory in major ways. One standard Aristotelian way of explaining friendship value is taking another's good to literally be part of one's own good. But Whiting objects that this is "colonizing" and domineering and thus intuitively unfriendly. Meanwhile, "brute" care, she argues, has a more basic explanatory disadvantage: it makes our choice of friends mysterious. To ground friendship in repeatable reasons traceable to friends' characteristics entails rejecting some intuitions associated with friendship about particularity and loyalty, such as the idea that any particular friend has some unique and irreplaceable value, or that one should be loyal to a friend who goes through a major change of character (which would disrupt the basis for the friendship). But this is a worthwhile tradeoff, she argues, for the explanatory advantages of a generic account of value (Whiting 1991).

The impersonal approach comes with some serious costs. Besides losing the above intuitions, it seems to yield a counter-intuitive account of how we select friends. Contingencies of accessibility, compatibility, and availability can perhaps explain why particular friends as individuals are especially valued, but these seem unappealingly like pricing and shipping considerations in shopping, where once one has decided what features one is looking for, one then selects based on availability and cost. Finally, by making the core valued features generic and repeatable, this seems to make friends fungible, freely interchangeable with others of their kind so long as they exemplify similarly virtuous character. In this impersonal account of friendship, to save the importance of reasons, we do not value all of a friend's features, only the good ones (at least not deeply and not with any moral significance). We thus lose the partiality and irreplaceability that seem characteristic of friendship. But where friends are valued for their virtue, then a more virtuous person would be more valuable, and one is bound to current friends only by pragmatic considerations, things like accessibility of virtuous people or personal preference (if the most virtuous person in town wears a perfume you can't stand, it will be difficult for you to spend time together). It may even seem desirable to "trade up" in one's friendships whenever possible. These counter-intuitive consequences are high theoretical costs. This failure of repeatable reasons grounded in friends' characteristics will have consequences for emerging social technologies. For example, it tends to count against the idea that companionate robots should simply be those that display characteristics deemed optimal for companionship.

One can try to preserve more intuitions about friendship while making space for generic reasons by thinking someone's character sets the stage for friendship without fully determining it. Laurence Thomas argues that this captures our sense that friendship involves choice, but also the idea that a friendship can spring up unexpectedly. Although he thinks one cannot shop for friends, he takes the fact that friendship can surprise us as evidence that friendship is not something in which people can arbitrarily decide to engage. Rather, some people's features make them well-suited to be our friends, relative to many other people out there, even though more than one person can possess relational properties of "fit" (compatibility with one's own preferences, values, and interests) that make a person particularly well-suited to be friends with you. These might include virtuous character traits, like loyalty, honesty, and generosity, but also less morally loaded features, including hobbies, interests, tastes, sense of humor, whether one is a morning person or a night owl, and so on.

Relevant features will enhance one's ability to engage with another in minimally structured interactions, absent the restrictions of clearly defined social roles. In such interactions, each is at the mercy of the other's judgment about what is helpful, enjoyable, harmful, and so forth. Minimally structured interaction characteristic of friendship requires a great deal of mutual trust. This, then, explains the sense of surprise and discovery that accompanies the development of friendships: we discover traits about these people that make them well-suited to be friends with *us*. Though the traits are repeatable, they are valued for relational reasons (the way their character and concerns "fit" with ours) rather than intrinsically, as on Plato's and Whiting's accounts. We can see the importance of appropriate "fit" when considering, for example, the frustration that can accompany discovering that a Facebook connection's idea of shareable content and appropriate commenting norms differ widely from one's own. These traits are background conditions for friendship, argues Thomas, because

> minimally structured interaction will be harmonious only if the parties involved are sufficiently attuned to the way in which each other views and interacts with the world. . . . Successful minimally structured interaction requires a shared conception of the good".
>
> (Thomas 1987, 220)

What these theories have in common is the importance they ascribe to repeatable traits. They all emphasize the importance of reasons in explaining how we select friends. Without certain repeatable traits, we would not befriend the people we do, nor choose to end friendships that turn out to lack these traits. Some such approaches hold that we value people only insofar as they instantiate a valuable property. Individuals are then fungible, a counter-intuitive result for a theory of friendship. The strategy pursued by Thomas seems problematic in a different way: friends' character traits are valuable because the right ones allow us to let our guard down, but both sorts of approaches fail to explain how a particular friend is valuable. The intuition that some unique person is valuable, by contrast, motivates a different approach to friendship valuing: one that prioritizes nonrepeatable features, thus avoiding the theoretical cost of fungibility.

4. Considerations in Favor of Particularism

The fact that a friend doesn't seem fungible would be easily explained if there is something uniquely valuable about that person. It would also explain the intuition that the end of a friendship is a genuine and significant loss, even if you have or can make other friends, and even if you are better off without this person in your life. These are the intuitions we saw emphasized in the case of Particular Patrick. It can also explain why your values are, and ought to be, shaped by your friends, as we saw in the case of Partial Padma. If something is both distinctive and valuable about your friend's perspective on the world, this can justify modifying your values in light of a friendship.

One such account of the intrinsic value of particular individuals is developed by Neera Badhwar, in a theory of what she calls "end friendship". Badhwar argues that end friendship is marked by "necessary irreplaceability" of the friend (Badhwar 1991, 484). "In an end friendship", she explains, "one loves the friend as an essential part of one's system of ends and not solely, or even primarily, as a means to an independent end. . . . In such love, one loves the friend for the person that she is" (1991, 483). End friends are constitutive goods, not necessarily maximizing goods, she claims. It seems plausible that one could believe, of a friend, "I might well have been happier on the whole without this friendship, whose presence is now a unique and irreplaceable constituent of my good" (1991, 492). This constitutive role amongst our ends explains the intuition that ending a friendship is a significant loss, even when it seems the best course of action.

Friends love what is metaphysically distinctive about an individual, which is, Badhwar claims, "her essential rather than incidental features". Essential properties, she says, "include both her character traits . . . and her unique perspective on herself and others: *her* view of the important and unimportant, *her* interest in herself and others" (1991, 483). These traits are not supposed to be shared with others, because friendship is not fungible: "in end friendship the friend cannot be replaced by another, for no other can have her essential features" (1991, 483).

We need a more robust account of essential features, however, if they are to serve such an explanatory role. Because "no other can have her essential features", that makes virtuous character, and repeatable character traits generally, insufficient for valuing a friend as a friend. But what properties could play the right role? Many of what we ordinarily take to be essential features of a person could be shared, and many unique features do not seem essential to the individual: fingerprints, for example.

4.1. *A Challenge to the Irreplaceability of Essential Features*

While the hope was that loving a person's essential properties—those without which they would not be the same person—would ensure their irreplaceability, taking this strategy may invite the problems of a repeatable-property theory. Nussbaum argues that one's "character and value commitments (as opposed to superficial pleasantness or advantageousness) are what each person is *kath' hauto*, in virtue of himself or herself" (Nussbaum 1997, 12), and such character and value commitments are repeatable. Plausible unique and nonrepeatable properties are often not central properties of a person; they tend to be relational or historical, or, at best, accidental. By focusing on what is unique to the individual, one risks missing out on a large part of what makes the individual who he or she most truly is.

Nussbaum points out that the relationship between repeatable character traits and valued individuals is open to a Euthyphronic contrast. Given that the list of traits one values and the traits of one's friends overlap heavily, one can ask which comes (logically) first. She poses a thought experiment in which a person becomes dear to her who lacks some of the traits on her list but possesses others: "If I loved him I'd change the list", she concludes (1997, 14). One possible explanation for this is that friends can reveal to us previously unsuspected values. Another is that our values change as a result of our friendships.

5. Partiality in Valuing

5.1. Epistemic Bias as a Norm of Friendship?

The first explanation runs the risk of being ad hoc. The second possibility suggests a different strategy. Perhaps we simply evaluate our friends' properties differently than other people's (Stroud 2006; Keller 2004). This is another way of handling cases like that of Partial Padma. Stroud offers an example in support of this thesis. Upon seeing or hearing about someone's truthful but potentially hurtful remark, one's response might normally be to think "what a jerk!" or at least that the speaker was somewhat tactless. The same remark made by a friend, however, would be judged as indicative of forthrightness.

Although taking this to be a norm of friendship could explain why people in friendships consider friends irreplaceable, the approach has costs. It leaves it mysterious why we become friends with some people and not others, and also leaves us with little resource in explaining why people might choose *not* to continue some friendships, where (as I will discuss in Section 9) repeatable character traits and values seem to play an important explanatory role.

5.2. Room for Partial and Impartial Valuing?

A more nuanced strategy for explaining friendship would accommodate both flexibility and more rigid principled valuing, but this approach ultimately faces problems as well. I will use Amélie O. Rorty's pluralist account to demonstrate the problems with such an approach. Rorty claims there is tension between valuing traits a person happens to possess, and valuing traits and properties because they belong to a person one values. She diagnoses this as a tension between what she calls flexibility, openness to changing one's values in light of attachment to a person, and rigidity, which involves staying true to one's (current) values. Given a scenario where one's values and a friend's behavior diverge, flexibility and rigidity suggest different strategies. Qua friend, one may interpret a sharp remark charitably. As a truth-seeker committed to one's values, one reevaluates the friend's character in light of the sharp remark. Rorty cautions that rigidity and its implied detachment can spell trouble, both for friendship and for one's own well-being. But the more flexible attitude is not guaranteed to go well, either. She notes that it can lead to shared psychosis, as well as the usual problems associated with wishful thinking and bias. There is no formula for deciding which to prioritize: "It is only in the details of [a] particular situation that [one] can determine what would be rational, what would be appropriate, what would constitute (whose?) thriving", she concludes (Rorty 1993, 88).

While the suggestion that flourishing may help to arbitrate between competing concerns is promising, I think a more explanatorily rich account is available by developing a theory of what is involved in a friendship's thriving, which does not leave important values in tension, as Rorty's does. Moreover, I think we have positive reason to reject the flexible form of evaluation she identifies as characteristic of the best friendships.

5.3. Against Epistemic Bias in Friendship

I do not doubt that favorable bias toward friends' traits, simply in virtue of their belonging to friends, accurately describes many people's actual judgments. But I question whether doing so is any more normatively valuable than people's judging themselves in an excessively positive light. As Stroud points out, the temptation to view friends in a special light parallels the ways in which people tend to cut themselves slack. If Aristotle is right that friends are other selves, self-deceiving about friends is no more plausible as a norm than self-deceiving about oneself would be, and if it is important to know thyself, it seems equally important to know thy friends. Evaluative bias thus seems an unsatisfactory way to capture the intuitions about particularity that motivate theories of friendship.

I think, however, that we can take one friendship norm to say we should look for the best in ourselves and our friends, more so than we do in others. Rather than doing so to deceive ourselves, it can help us to identify and encourage latent strengths. Neera Badhwar offers one such account (Badhwar 2005). Friendly judgment of a blunt remark could be explained by saying that one recognizes that one's friend is tactless in making that remark, but at the same time wishes to draw out the friend's potential for forthrightness, which requires fearlessness about speaking the truth, which the friend displays. It would be inappropriate and probably paternalistic to take the same approach with casual acquaintances, even with the intention to help the person improve. (For more extensive discussion of this issue, see Chapter 3.) But it seems a valuable part of friendship to see potential in a friend even when he falls short, so that one might say, "you hurt his feelings when you said that, but I think it's important to speak the truth . . . maybe you should find a more tactful way to make your point next time".

This sort of account would capture intuitions about friendship-appropriate beliefs, without advocating the more radical conclusion that truth norms conflict with norms of friendship. However, explanation of friendship value is then still needed.

6. A New Strategy

It would be helpful to have an account of what is involved in valuing friends over time and through changes that does not explain this value as a violation of epistemic norms. And as shown in Sections 3 and 4, identifying friends with either their repeatable or nonrepeatable properties seems problematic. Instead, I think that we should stop focusing on traits of individuals, and start looking at the bigger picture. Instead of looking at individual traits of friends (or even individual friends) as sources of reasons, it is helpful to think of friendships as composite objects with friends as parts. These composite objects can be taken to be valuable in ways that provide reasons with more explanatory power than previous accounts.

I propose that we conceive of friendships as composite objects composed of individual people as parts, much as organisms are composed of individual organs but not reducible to them. To compose an organism, the individual components must be appropriately interrelated and interresponsive: a heap of parts is not a creature. Likewise for friendships: friends must be interrelated and interresponsive in order to compose a friendship, and for the friendship to persist through time, their interdependence must also persist.

Thinking of friendships as being like organisms, composed of interdependent and interconnected organs (in this case, friends) explains why historical and relational properties matter: sustained patterns of interaction, emotional interdependence, and mutual influence are what constitute people's jointly composing a friendship, distinguishing them from mere collections.

Furthermore, identity conditions of friendships may be tied to their component parts: just as an axe that has had major parts, such as the axe head or handle replaced, could be plausibly said to no longer be the same axe, the irreplaceability of friends may be explained by positing that replacing friends destroys the identity of the valued object: the friendship.

In addition, taking friendships to be not just composite objects, but relevantly like organisms in that they can be capable of flourishing and withering, provides additional resources for understanding reasoning in friendship. The well-being of such a whole requires both the well-being of its component parts, and the successful interaction of those parts.

In friendship, it can matter not just that friends value the same things, but that they value the same things in virtue of their being caused by their past association. Thus, even some apparently repeatable characteristics (such as sharing values) end up having historical and irreplaceable dimensions. But although including relational and historical properties as reason-giving seems to help secure particularity, these are not, intuitively, central features of a person, as Nussbaum pointed out, and many seem intuitively as though they could vary while leaving the individual substantially the same person.

In some respects what I propose here resembles that proposed in Kolodny (2003). However, my account differs from his on some key points. I aim to give an account of friendship, while he intends for his account to also cover parents' love for children and other cases of familial love, in which one has substantially less choice about one's object of love. This introduces problems. As Helm (2010) points out, for example, Kolodny has a difficult time giving sufficient grounds for *ending* a loving relationship, besides that other concerns may trump those reasons provided by history and (non-reciprocal) relational properties (Helm 2010), while as I argue in Section 9, my account provides a theory of how friendship itself can give reasons to end a relationship. This in part stems from my claim that the relationship is a composite object with friends as parts, so that valuing the composite is (also) valuing its parts, while Kolodny holds that love involves *both* valuing the individual and the relationship (which he does not construe as an object), and occasionally takes these to be two separate things, which I explicitly reject. Finally, because I take relationships to be not only objects, but living objects capable of flourishing or withering, whose well-being is partially dependent upon the well-being of constituent parts (the friends), I have some explanatory resources available (particularly for the role that character plays in friendship—as presented in sections 7 and 9) that are not part of Kolodny's account, nor is it clear how they could be incorporated into such a view.

One might wonder, why not just specify a list of things one values about friends, a list that includes both repeatable character traits, interests, and values, as well as historical and relational features? My response is to appeal to explanatory depth. Even if a list and a structure can explain all of the same things, the structure has more to say about why *those* items and not some others, and how they relate to one another. Explaining friendship as an organism with friends as interresponsive parts yields a complex of both individual (human), historical, and relational properties (those constitutive of their interdependence) which makes sense not only of *what* is valued as a list of individual reasons, but how those reasons hang together.

We commonly say of good friends that they value their friendship, and I take this to be literally true. What friends value includes persistent patterns of interaction between particular individuals. Taking the friendship, not just the friend, to be an object of value captures another intuition about friendship: it unfolds over and through time, and is not the sort of thing that, even in principle, one could enjoy without some upfront investment. This can explain and incorporate intuitions about the importance of history, reciprocity, and relational properties. These are not necessarily important to the individual's identity, but constitutive of the friendship. This makes an important role for shared history, as a constitutive feature of an object of friendship value.

We also get a way to capture the particularity of friendship valuing without taking it to involve epistemic bias, and also a way to flesh out Badhwar's appeal to "essential properties". The properties we value are not essential to the friend qua human being. Rather, they are constitutive properties of the friendship as a persisting entity that includes both friends.

Other kinds of social relationships—from business relationships to neighbors and families—can persist whether or not the participants are emotionally interdependent. One can still be a client, lawyer, customer, shopkeeper, employer, employee, spouse, parent, student, or teacher even if one's emotional states are not affected by those to whom one is related as these relational terms imply, nor does the other party's emotional state depend upon one's own, as a constitutive feature of their being one's lawyer, client, vendor, customer, employee, employer, spouse, child, teacher, or student. The exception to this might be enemies, about whom it seems best to say that their emotional interdependence is not valued by the participants, or not conducive to their flourishing. While it is difficult, sometimes, to sort out non-literal normative appraisals disguised as metaphysical claims from literal claims about what it takes to constitute a friendship, it seems inapt to call people friends if they are not emotionally interdependent. Thus, friendships seem to necessarily include historical and relational features such as "happy-now-because-my-friend-got-a-promotion" that capture this emotional interdependence. (Recall that friendship can overlap with the other sorts of relationship sketched above—my sibling can also be my friend—and to the degree that we find emotional interdependence in these other relationships, this may be taken to indicate that these relationships are also friendships.)

7. Reconciling Apparently Contradictory Intuitions

I have proposed that friendship value is characterized by taking its object to include but not be limited to the friend; rather, it extends to cover the friendship and oneself as part of it. Taking this approach allows us to explain many important intuitions about friendship, without positing irreconcilable tensions between intuitions or values.

The intuitive importance of reciprocity and mutual responsiveness suggest a connection between initial fit amongst friends, the value of historical properties, and sensitivity to another person's values. If one object of value is the friendship, a composite object whose parts' composing the whole is characterized by such reciprocity and responsiveness, then we are—to some extent—asking the wrong question about friends when we ask, what about that person guarantees that I will (or ought to) value him, come what may? This is a problem for parents, perhaps, but not for friends, because reciprocity seems necessary for friendship to take place. People who do not reciprocate one's care are not properly objects of friendship value because they are not friends, properly considered.

This approach is consistent with Badhwar's point that friends are constitutive rather than maximizing goods in our lives, while explaining it differently: if what we value is participating with our friends in a reciprocal interdependent arrangement constituting a composite object (a friendship) over time, in which we adjust and attune to each other's values, then we will to that extent be inclined to consider our friends' values and perspectives both intrinsic to them, and valuable to us, as part of this friendship. Yet, this is not simply to value them in virtue of instantiating some more generally valuable characteristic (like courage or forthrightness) but in virtue of their role in our lives, as irreplaceable constituents of something we value. The friends are irreplaceable at least partly in virtue of the identity conditions of friendships, which do not seem to survive the replacement of friends.

Character traits and repeatable properties remain important in several ways, without entailing conflict between repeatable and nonrepeatable properties. The appearance of conflict arises from confusion about when and where each should be a predominant concern, but this is avoided by taking them to play different roles in a valued friendship. One way repeatable properties may be important comes at the stage when we select our friends. Some repeatable traits are necessary for someone to be a desirable part of this complex organism, partly, as Thomas notes, as necessary preconditions which allow us to trust people without relying on the external constraints and protections of a more structured social role. But this does not mean that every person who meets these conditions should become a friend; some people may strike us as fine and admirable individuals without leading us to desire to enter into a friendship with them.

Once a person has enough of these traits that we are willing to let down our guards, friendship can commence, provided both people are interested in pursuing it. Given time and reciprocation, each person's judgments may eventually be taken on by the other as being of brute value. But this brute value is not as much of an explanatory disadvantage as Whiting thinks. Having the friends one does helps explain why one has the values one has. Rather than think that either one's friends or one's own values are privileged *in an established friendship*, either is open to adjustment and pruning, as part of continuing involvement in a friendship. But this flexibility need not mean that we set our own values aside as unimportant. Rather, friendship is characterized by the fact that both friends' values are taken to be considerations for each friend, because friends are both parts of the valued friendship, and because coordinating interests and values is necessary for people to compose a friendship. Friends also have reason to compromise and curtail competition between their values. They value themselves and their friends as parts of the friendship, and they want the friendship to flourish.

This does not make values merely of instrumental value in ensuring fit, however, and the way that people's value commitments relate to friendship assures the continuing importance of friends' character. One's 'vision of the good' matters because, among other things, it directs a person's actions, and when one cares about a person (oneself or another), one wants for that person what is (in fact) best. Bad values will direct a person at bad, harmful ends. One wants to get one's values right, and to be connected to people whose values are right, because rather than despite the fact that they will affect one's own values. This is an issue that will be addressed in greater detail in Chapter 3.

Sensitivities to each other's values can produce unhappiness in the form of shared delusion or reinforced vices (Rorty 1993). But when one object of value is the friendship, which encompasses both friends, then coordinating one's values with the friend is seen as part of advancing the interests and promoting the well-being of the friendship. The fact that we desire the flourishing of the friendship helps determine which values it would be best for the friends to have. Because the friends are constitutive parts of the friendship, their individual well-being is necessary for it to flourish, just as the constitutive parts of an organism must be healthy in order for an organism to flourish. Each friend's well-being, then, partly constitutes the well being of the friendship, so their concerns are already incorporated as part of one's concern for the friendship. But the value of the friendship comes before the value of virtue in the order of explanation in an established friendship. We want friends to have certain repeatable traits because we care about them and want them to do well, and want them to be part of a flourishing friendship with us; their virtue does not fully explain why we value them. Thus, valuing virtues in this way does not imply fungibility of friends, as it is not the case that any other equally virtuous person would be equally valued.

On this account, then, if a person sets aside his interests in the interest of the friendship (as when someone falsely says something is "no big deal", to keep the peace), he harms a part of the friendship and thus fails to promote its well-being. This is consistent with the thought that self-abnegation is not a feature of a good friendship, and that friends should value both themselves and each other. This does not, however, mean that one can *never* set aside one's own interests to help a friend; more on this in section 9.

Although Whiting worries that taking a friend's good to be part of one's own good is colonizing, taking *both* friends to be parts of the valued friendship alleviates this concern. Where both are equally important to the friendship, neither's concerns take automatic priority. In valuing the friendship, neither person's good should be seen as subsuming the other's.

A friend's virtues play an explanatory role in our valuing, but are themselves also explained in terms of other things we value, including the friend's well-being and the flourishing of the friendship. Thus, the connection between valuing a virtue and valuing a friend depends upon the status of the friendship. Initially, a potential friend's values are checked against a potentially revisable list of one's own values. One can be delighted to find a person who shares them. This makes such traits necessary features of potential friends, without making them sufficient for friendship. But the replaceability problem drops out once a friendship is established, because friends' values become open to revision as the friendship unfolds, as friends reciprocally respond to each other's interests and perspectives. Particularity is thus guaranteed by friends' historical properties and combinations of traits being properly related with one's own, so as to constitute a flourishing friendship. Historical and relational properties are thus crucial constituents of a friendship, and some personal trait's bearing the right historical and/or relational features is a prerequisite to friends' (ideally) responding to that trait as a friend would. This underscores the importance of getting clear on what social media does to friendship, as interacting through social media has become a commonplace way for people to interact and influence each other.

Finally, taking a friendship to be an object of friends' value naturally suggests a role for the intuition that we judge friends differently than non-friends. As a friend, one plays a special role in highlighting potential and helping to direct a friend's character, which may warrant focusing on different aspects of a person's action than would be justified were the person not one's friend. Insofar as one is concerned with both one's own and one's friend's flourishing as parts of the flourishing of the friendship, one has reason to see oneself and one's friend both clearly and positively. To the extent that seeing potential goodness in a friend is pragmatically beneficial for promoting the friend's well-being and helping the friend become the best person possible, one ought to see the best in one's friend, and be open to the friend returning the favor. This is not a violation of epistemic norms but the result of additional pragmatic norms that come into play given one's interest in the flourishing of the friendship, which requires both veridical and helpful observation of the parts and the whole.

8. Historical Aspects of Friendship

I do not claim that history is valuable, *simpliciter*, but that it is partly constitutive of friendship, and that in its capacity as a component of friendship, history is thereby desirable: one wants one's friendships to persist (all else being equal, with the caveats dealt with in the section on ending friendships), and one values one's old friends partly (but not solely) for the duration of the friendship. Although this will strike some as intuitive, it may raise questions for others, and so I will say a little bit more about my conception of the role of history in friendship.

In some cases, we do not expect friendships to last very long—friendships between young children, for instance. Although one might be happy to grant that there are no instantaneous friendships, it still might seem as though one can have a perfectly good (in some sense) friendship without it persisting very long. However, such ephemeral friendships seem not to represent an ideal we aim for in our friendships, but rather a resignation to facts about the fickleness of youth. Furthermore, when such friendships do persist well beyond their expected duration, this seems to be a valuable thing to cite about them. Saying that one has been friends with someone since childhood seems to be a way of saying something quite positive and valuable about the relationship.

9. On Ending Friendships

I close by considering an application of my theory to a situation where valuing both repeatable and nonrepeatable properties seems important: when a friendship ends. It seems both that ending a friendship can be a good and reasonable decision, given certain events or character traits or circumstances, and at the same time, that even a reasonable and advisable ending of a friendship seems a genuine and significant—perhaps even tragic—loss.

If the friendship is an object of friendship value, friends ought to realize that when a friendship's flourishing is impossible, its concerns are tragically unsatisfiable. That still leaves the well-being of each component part (each friend) at issue. Friends can (and should, in such cases) choose to end a friendship for each other's sake, or their own, as a result of valuing themselves and each other. As Socrates says in the *Symposium* (Plato 2009), we ought to value goodness over wholeness. In his graphic example, we are willing to amputate even our own limbs when they become gangrenous. One might choose to sever ties with a friend because one recognizes that one's own or one's friend's problems are unfixable and/or likely to contribute to the other's unhappiness. This is not inconsistent with also valuing the whole, the friendship: it just says it is not the only concern, but this was never the case to begin with. What my theory predicts is that even in a case like this, the end of the friendship will often be viewed as a tragic loss.

Loss of even a bad friend is construed as a loss of a constitutive rather than a maximizing good, where the friend is a necessary part of the friendship, itself a constitutive good considered more broadly, as an important feature of a good life. A theory based exclusively on valuing people as instantiations of valuable properties seems to make the loss of a bad friend without cost, because on such an account a bad friend lacks valuable virtuous character, or because they can be (or are in fact) replaced by another friend with better character. But this seems too harsh a conclusion to be compatible with friendship value.

Friendships can be ended with good reason, but even where friends are bad, something valuable seems to be lost when the friendship ends. As Aristotle notes, the dissolution of a friendship may reasonably be due to the deteriorating character of a formerly good friend, but this does not make the choice to end a friendship an easy one, as we would expect if friendship were purely about accruing values of which friends happen to be the bearers.

"Should the friendship be dissolved at once [as soon as the friend becomes bad]?" he asks. "Surely not", he replies. "If someone can be set right, we should try harder to rescue his character than his property, insofar as character is both better and more proper to friendship". But if the person is "incurably vicious" and cannot be set right, then "the friend who dissolves the friendship seems to be doing nothing absurd" (Aristotle 1999, 141, 1165b). Rather than take this to be a case where self-interest and the friend's interest are in competition and one's own interests trump, this looks like a case where it is advisable to cut one's losses. All things considered, it would be better to be both good and whole, but given a hard choice, where it is impossible to have both, we salvage what we can from amongst what we value.

Because I have said that even the loss of a bad friendship is a loss, one might wonder whether this is always so, or whether some friendships are such that their loss is not worth regretting. Suppose one has had quite a vicious friendship, based on shared values one now (fortunately) recognizes as mistaken and so rejects.

To put some flesh on this suggestion, imagine a friendship between Adam and Barney. Both Adam and Barney, when they meet, are highly misogynist, and they bond over jokes about women drivers and catcalling pedestrians. They reinforce each other's misogynist tendencies, becoming worse by association with each other over time. Adam eventually comes to believe that his prior values and behavior are terribly mistaken, and sets out to reform himself, among other things by avoiding association with his old friend, to avoid lapsing into old habits. He may now come to regret 1) having been the kind of person who could have become friends with Barney and 2) having had a friendship with Barney that made it more difficult for him to stop being such a person.

Am I committed to saying that this constitutes a loss for Adam?

Yes and no. Adam might regret that the relationship was so flawed it had to be terminated, and wish that Barney had been redeemable. This need not tip the scales in favor of reviving the relationship, nor need it lead to nostalgia for old times, if one is genuinely happy to leave that life behind. The apparent value, to Adam, of his relationship with Barney itself helps explain why Adam feels the need to distance himself—if he were indifferent, Barney might be less of a temptation to lapse into old ways. But it is compatible with holding that friendships are valuable to say not every friendship is worth valuing, particularly when it fails on some important dimension of friendship by one's own lights. In this case, Adam judges that he is better off without the influence on his character that Barney introduced, and thus that his friendship was, in a sense, harmful to him. Good friendships do not harm those involved, and good friends do not drag each other down, so the friendship fails as a friendship on Adam's own terms.

10. Conclusion

In my account of friendship, friends are constitutive components of friendships. I showed how such an account, in which friendships are objects of value to the friends, allows valuing repeatable and nonrepeatable properties to both play important but different roles in the development, sustenance, and dissolution of friendships, and showed how epistemic bias's apparent importance can be accommodated without putting truth and friendship at odds. This theory thus explains important intuitions as different but coherent features of friendship.

If friendships are entities with friends as parts, and its flourishing is a central concern of friends, then each kind of property ends up being assigned a role, and their apparent conflict is the result of misapplying concerns about one aspect to another aspect of friendship. Partiality is explained as a combination of concern for the friend's well-being and a feature of friends' responsiveness to each other, necessary for the long-term coordination characteristic of friendships. Particularity follows by taking friendships to be composite objects with historical and relational properties as necessary constitutive features: we value them and so think friends are not (and could not be) interchangeable, not by finding some mysterious and irreplaceable essence unique to each friend but in virtue of their role in a friendship. Repeatable traits, such as virtues and shared conceptions of the good, are important preconditions of friendship in allowing potential friends to become coordinated, but also valued for their role in sustaining friendship: we want our friendships to flourish, and where friends are vicious, friendships cannot thrive. If friendships cannot thrive, then we are justified in ending them, even when we still care for the people with whom we've been friends.

Now that we have seen some of the explanatory work that can be accomplished by positing that friendships are compound entities with friends as parts, it will be helpful to consider in more detail of what exactly these entities consist. This is the focus of the next chapter.

Bibliography

Aristotle. 1999. *Nicomachean Ethics*. 2nd ed. Translated by Terence Irwin. Indianapolis, IN: Hackett Publishing Company. https://books.google.com/books?id=-Mf5XV8q6CgC&dq.

Badhwar, Neera Kapur. 1991. "Why It Is Wrong to Be Always Guided by the Best: Consequentialism and Friendship." *Ethics* 101 (3): 483–504. doi:10.1086/293313.

———. 2005. "Love." In *The Oxford Handbook of Practical Ethics*, edited by Hugh LaFollette, 42–69. Oxford: Oxford University Press. https://books.google.com/books?id=emstfpfaT2wC.

Cooper, John. 1980. "Aristode on Friendship." In *Essays on Aristotle's Ethics*, edited by Amélie O. Rorty, 301–40. Major Thinkers Series. University of California Press. https://books.google.com/books?id=QHYKYZGKTgwC.

Helm, Bennett. W. 2010. *Love, Friendship, and the Self: Intimacy, Identification, and the Social Nature of Persons*. Oxford: Oxford University Press. https://books.google.com/books?id=8s0UDAAAQBAJ.

Keller, Simon. 2004. "Friendship and Belief." *Philosophical Papers* 33 (3): 329–51. doi:10.1080/05568640409485146.

Kolodny, Niko. 2003. "Love as Valuing a Relationship." *The Philosophical Review* 112 (2): 135–89. doi:10.1215/00318108-112-2-135.

Levine, Michael. 1999. "Loving Individuals for Their Properties: Or, What Was the Colour of Yeats' Mother's Hair?" *Iyyun, The Jerusalem Philosophical Quarterly*, no. 48 (July): 251–67.

Nussbaum, Martha. 1997. "Love and the Individual: Romantic Rightness and Platonic Aspiration." In *Love Analyzed*, edited by Robert E. Lamb, 1–22. Boulder, CO: Westview Press. https://books.google.com/books?id=-PVZ1q-_KFkC.

Plato. 2009. *Symposium*. Translated by Robin Waterfield. Oxford World's Classics, Oxford University Press. https://global.oup.com/academic/product/symposium-9780199540198?q=9780199540198&lang=en&cc=us#.

Rorty, Amélie O. 1993. "The Historicity of Psychological Attitudes: Love Is Not Love Which Alters Not When It Alteration Finds." In *Friendship: A Philosophical Reader*, edited by Neera Kapur Badhwar, 73–88. Ithaca, NY: Cornell University Press. https://books.google.com/books?id=YBCI22Lz_I0C.

Stroud, Sarah. 2006. "Epistemic Partiality in Friendship." *Ethics* 116 (3): 498–524. doi:10.1086/500337.

Thomas, Laurence. 1987. "Friendship." *Synthese* 72 (2): 217–36. doi:10.1007/BF00413639.

Whiting, Jennifer E. 1991. "Impersonal Friends." *The Monist* 74 (1): 3–29. doi:10.2307/27903221.

2 What Shared Identity Means in Friendship

1. Introduction

In the previous chapter, I argued that thinking of friendships as compound entities with friends as parts offers explanatory advantages helping to make sense of how friends balance concern for repeatable qualities like character traits with unique characteristics like individual and shared history and personality quirks. Here, I take up the nature of these entities in more detail. By delineating what it takes to constitute a friendship, we can make headway both in sorting out the status of friend-like technologies such as social robots, and in evaluating how well communication technologies like social media can support friendships. In addition, we can gain a better understanding of another commonplace but underdeveloped feature of many people's experience of friendship: its impact on individuals' identities.

My starting point is a widely shared intuition: close friends seem to somehow share identity. In Aristotle's work, he frequently claims that friends are "other selves". But this idea is not restricted to Aristotle. A survey of the literature on friendship reveals that it is often described as a union, or explained in terms that imply that friends share identity. Here are a few examples of the phenomenon, taken from a wide range of accounts of friendship: Augustine describes himself and a friend as "one soul in two bodies" (Augustine 1993, 56). Gilbert explains close relationships such as marriage and friendship in terms of "fusion" (Gilbert 1990). Telfer says friendship involves a "sense of a bond" between friends (Telfer 1970, 226). Schoeman describes friendship as "a way of being and acting in virtue of being united with another" (Schoeman 1985, 281). Friedman describes friendship as a federation of individuals (Friedman 1998). White says friendship involves solidarity (White 2001). Sherman says friendship involves "sameness of mind" and "extension of self", as well as a "relaxing of boundaries" and "sense of union" (Sherman 1987, 282). Hampton describes close relationships as ones where "there is an intimate connection between the parties—to the point where the pleasures of each are advanced when the other's needs and desires are satisfied"(Hampton 1993b, 158).

All things being equal, it would be desirable to explain the widespread appeal of this idea by supposing that it is true. But there are a number of serious objections to this thesis that would need to be overcome in order to make it plausible. In what follows, I will present one way to do so. I suggest that the objections disappear, or at least are substantially blunted, when one considers carefully what is involved in ordinary claims about identity.

The solution, as I see it, is found in the realm of social ontology. Thinking of friendships as composite social objects with friends as parts can explain appearances of identity without the theoretical costs of previous union accounts of friendship. In particular, resources from contemporary work in metaphysics on the composition of complex objects can both enrich our understanding of social ontology, and explain a curious but powerful intuition about friendship.

The most common way to unpack this intuitive notion of unifying shared identity in friendship is in terms of similarity. But it is troubling to suppose that friendship depends upon similarity of friends (Williams 1981; Cocking and Kennett 1998). Differing perspectives and complementary strengths are among the potential benefits of friendship, and a good theory of friendship should not rule out such differences. Friendship's union is also sometimes taken to involve making a friend part of oneself or part of one's life. Although this would help explain shared identity, it is inappropriately colonizing or domineering (Whiting 1991). A good account of friendship must avoid a foundation that too closely resembles the colonizing or domineering found in abusive marriages, pushy stage parents, and other unfriendly paradigms of relationship. In general, union accounts are counter-intuitive insofar as they fail to recognize boundaries between friends. Bennett Helm, for example, criticizes union accounts of love and friendship on the grounds that

> appeal to union seems excessive in that (a) it questionably undermines one's autonomy and distinctness from one's beloved and (b) it unsatisfactorily construes intimacy in egocentric terms and thereby fails to make sense of the idea that in love we have a concern for our beloved for *his* sake.
>
> (Helm 2010, 147)

To be plausible, union accounts then cannot dissolve boundaries. But it is, at first blush, hard to see how such an account could still be a union account.

This makes unifying accounts less appealing. Some authors qualify their claims by rejecting *literal* or metaphysical union, as when Sherman (1993, 282) says in a footnote that "In making these remarks [about friendship as union], I hope to be steering clear of the thicket of metaphysical issues having to do with personal identity and the constitution of the self". But this leaves it mysterious just what this union is, and why it has such widespread appeal.

Taking friendship to be a unity is attractive, but also threatens to endanger individuality and differences between friends. Aristotle appears to attempt to avoid this by saying that friends are "other selves" (for example, in the *Nicomachean Ethics*, (Aristotle 1999, 1161b, 1166a, and 1170b)), but this just sounds contradictory upon reflection. If something is other, it seems impossible that it be oneself, and if something is oneself, it does not seem to be other. One could bite the bullet and say that our intuitions mislead us; friendship is not genuine unification. But it would be preferable to explain how and why it appears to unify, and desirable to resolve the tension so that Aristotle's term is literally true.

I argue that we can reconcile unification with difference in friendship by taking people who are friends with each other to be parts of a whole, a friendship, and who thus share identity via this whole. It is no accident, I argue, that Aristotle called such friends "other selves": considerations of identity can explain how such friends care for each other. Shared identity is often left at the level of metaphor. But the intuitive appeal of shared identity deserves deeper exploration. Investigation of identity between parts and the wholes they compose captures these intuitions and reveals connections between important features of friendship.

This is not the only way to resolve the apparent tension between the self/other dichotomy and friends' experience of shared identity. Relational theories of identity, for example, claim that the self is constituted by social relations, undermining the traditional dichotomy between self and other. But the approach I propose here has the advantage of retaining the intuitive distinction and recognizing when it does not apply. It allows us to talk about the self as both an individual in contrast to others (whether intimates or strangers) and as something shared with "other selves" versus the rest of the world. It thus offers rich explanatory resources, in addition to the advantages noted in Chapter 1. Despite this, one might worry that explaining friendship via metaphysics seems so odd or counter-intuitive that any theoretical benefits would be outweighed by the costs of appealing to metaphysics in the first place. But while I will be using a theory of identity to explain friendship, I will not insist that we become metaphysical realists of any particular sort to accept it. If theoretical commitments preclude adopting what I advocate as a precise and literal description of what there is, I believe that this approach also works as an account of what we value when we value friendship.

A few preliminary remarks about my methods are in order.

Shared identity is most plausible in very close friendships, as compared to mere casual acquaintances. This suggests compatibility with a roughly Aristotelian strategy for making sense of friendship. As we saw earlier, the "core" kind of friendship is something like what he characterized as virtue or character friendship, while more casual relationships based on mutual convenience or superficial pleasure in each other's company earn their title by resemblance to the paradigmatic case. I consider, therefore, examples of friendship that would fall into Aristotle's category of virtue friendship. One might worry that this would restrict the range of possible examples too much, since only the perfectly virtuous could have perfect virtue friendships, and such individuals are very rare. But it has been plausibly argued that we can understand Aristotle's theory to be connecting the quality of friends' character to the quality of their friendship (Cooper 1977). Mostly-virtuous people can have mostly-virtue friendships, halfway decent people can have halfway decent friendships, and so on. This relationship between virtue and friendship will be developed in much greater detail in the next chapter. But for now, it will be sufficient to take virtue friendship to describe an ideal of relationship that real life friendships more or less closely approximate.

This chapter is intended primarily as an analysis of friendship, not exegesis of Aristotle. However, insofar as I am sympathetic to an Aristotelian approach, the two tasks are to that degree interrelated. If one is unpersuaded by an Aristotelian theory of ethics or friendship, one may still find it valuable to explore an identity-based account of friendship. But because Aristotle makes other selves such a central part of his theory, much discussion of the role of identity in friendship has been framed in terms of his ideas, so I will say a few things about how this project coheres with his work.

Aristotle thinks that character matters, that character is integral to identity, such that valuing a person's character is valuing the person, and that the best of friends are other selves (Aristotle 1999). One way of reading this, indeed, a very natural one, is that the best friends thereby have very similar—or even the same—character. Cocking and Kennett, for example, offer an interpretation of Aristotelian friendship according to which virtue friends are mirrors of each other, and especially mirrors of each other's characters (Cocking and Kennett 1998). Whiting takes things one step further, arguing not merely that virtue/character friends are very similar, but that they exemplify the very same virtuous character traits, which are what is really valued in virtue friendship, on her account (Whiting 1991). In this, as we saw in the last chapter, she rejects the portion of Aristotle that holds that valuing a person's character is valuing the person, or at least that valuing the person does any interesting normative work. Thus, the work in which she defends this is titled "Impersonal Friends".

But when one thinks that virtue friends either must have very similar characters (in order to be mirrors of each other) or else the very same character traits/virtues, this seems to make differences of character a threat to virtue friendship. But that seems wrong, and perhaps even a little bit disturbing. Differences can be a valuable part of good friendship (Williams 1981). Complementary differences, in particular, seem as though they can greatly enrich friendships. One can see how the idea that difference is a threat to friendship gets going: after all, a liar and an honest person seem unlikely to have a successful friendship, and their differences appear to be relevant to explaining such a friendship's poor prospects. But a friendship between a cautious person and a more adventurous individual does not necessarily face the same kind of disadvantage, so the problem with the first case cannot merely be difference. This could mean different people can be close and virtuous friends, as long as they are both good, albeit in different ways. But then that seems to threaten the "other selves" portion of his theory, and that seems to be a powerful intuition to give up, not just an eccentricity of Aristotle's theory.

So, it would be desirable to reconcile difference (the "other" part of "other selves") with sameness: identity. Doing so lets us integrate Aristotelian character ethics with a plausible theory of friendship that is true to our intuitions and experiences, without running into contradictions or tensions that get explained away as mere metaphor. Friendship is an important human good, and it would be good to show that its underpinnings can be rational and consistent.

Shared identity is a quality of the ideal of friendship that can be more or less fully realized in particular real-life cases. It makes sense, then, to begin an explanation of the appearance of identity in friendship by considering how it coheres with other norms and ideals of friendship, especially close friendship of the sort characterized as virtue friendships.

In the first section, I argue that key features of friendship align with our usual reasons for thinking objects are parts of composite wholes. I then appeal to a theory of composition as identity to explain how parts of a whole can be at once the same thing, and yet differ from each other. I show how this theory resolves the apparently contradictory claim that friends are "other selves". This metaphysical account captures reasons for thinking friendship unifies, without threatening difference between individuals. In addition, it explains the value of differences among friends in terms of different strengths, which contribute to the flourishing of a friendship, an entity that encompasses both friends.

Note that my goal is to explain *how* friends can plausibly be thought of as unified, or as sharing identity. I do not aim, here, to explain *why* this is a desirable feature of friendship. (Analogously, as I noted in the last chapter, a mechanistic account of color perception can explain *how* the appearance of redness is produced in the visual systems of human beings, but not *why* one sees red as red rather than some other color.) But given the widespread appeal of such a claim, I think it is worthwhile to make it plausible. Reframing the issue, while not explaining why unification is desirable, may introduce constraints on what a reasons-why account could look like. It also suggests alternative ways to approach the question, by introducing friendships as objects of concern, capable of flourishing and withering, and whose well-being is intertwined with our well-being as people and which stand somewhere between narrowly egoist and selflessly altruistic concerns.

2. Friendships Show Symptoms of Parthood

Several features of friendship look like conditions under which one might reasonably take many differing things to be organs of a single organism. This suggests a way that friendship could involve unity without ruling out difference.

Different objects are sometimes considered organs of the same organism when they are mutually responsive to changes in each other's condition, especially across a wide range of circumstances. For example, different organs in the human body are mutually responsive to changes in each other's states, and this might partially underwrite their composing a whole human being. Additionally, different items are often taken to be parts of the same whole when they work together to collectively engage in some shared activity. The leaves, branches, and root system of a rosebush, for example, work together to draw energy from sunlight and soil in order to produce blossoms. Finally, different things can make up a unified whole when they are not just mutually responsive to changes, but actually shape each other's development. Each of these can independently give one reason to believe that one is confronted by parts of an organism which together compose a unified whole. Collectively, they provide stronger reason to believe this. Things that are mutually responsive to each other, *and* work together, *and* mutually shape each other, especially over extended periods of time and across a wide range of circumstances, seem especially good candidates for this kind of parthood.

First, consider mutual responsiveness. Friends value each other and value being in this reciprocal state of valuing: they value their friendship. Thus, they are mutually responsive to each other's interests and values. Suppose Alfred and Bruce are friends. If Bruce cares about vintage Corvettes, Alfred has some reason to care about them too, simply in virtue of Bruce's interest in them. This need not lead Alfred to acquire his own vintage car, but it does mean that seeing one on the street will remind him of Bruce, it might prompt him to point it out to his friend, he should express and experience delight at Bruce's new purchase, or sadness or unhappiness if something happens to Bruce's beloved vehicle. The two friends end up with coordinated interests at any given moment, as well as interests that remain coordinated over time. This feature of friendship has struck some as so important that they think it distinguishes friendship from more disinterested instances of care. Stocker argues that friends make each other's values and interests part of their background concerns out of which their own activities flow (Stocker 1981). This helps distinguish friendship from our more disinterested concern for acquaintances and strangers. The widespread popularity of computer-mediated communication technologies such as social media speaks to the importance people ascribe to living lives that allow them to be mutually responsive to their friends.

Note that this responsiveness is an addition, not a replacement, to the goings-on in one's life, via social media and in person. In becoming friends, people retain their own interests and values, but take on new ones as well, in virtue of the friendship. Their interests and values end up being coordinated via their mutual responsiveness to each other's concerns. This systematic coordination partially explains the intuition that friends unite or together compose a whole.

Next, consider shared activity, another feature of parthood. Friends paradigmatically value doing things together: restoring old bikes, working out together, sharing meals and movies, traveling together, and so forth. Aristotle observed that

> Whatever someone [regards as] his being, or the end for which he chooses to be alive, that is the activity he wishes to pursue in his friend's company. Hence some friends drink together, others play dice, while others do gymnastics and go hunting, or do philosophy. They spend their days together on whichever pursuit in life they like most; for since they want to live with their friends, they share the actions in which they find their common life.
>
> (Aristotle 1999, 1172A)

Although this might appear to imply that friendship simply consists in acting together, this is incorrect: it offers necessary rather than sufficient conditions for friendship. We share activities with co-workers, fellow committee members, and so on, without thereby becoming friends. What helps set friendship apart is that in friendship, not merely the activity but the sharing itself is valued. I may work together with a co-worker, even one I don't like much, because I want to get the project done and this person is a means to that end. But with my friends, it matters that this particular person share the activity with me, and in many cases, it matters more that we share *some* activity or other than that it be any particular pursuit (we may be equally content to go for a hike or to cook a meal together). If bike restoration is a component of our friendships, then we value fixing bikes with our friends. Fixing bikes, for close friends, would not merely be about working on a valuable project with convenient allies. To reiterate, shared activity is insufficient for friendship. As Nancy Sherman (1993) has argued, activities can be shared on a much smaller scale, from long-term projects such as raising a child to short-term activities like conversations. But it looks like no accident that these sorts of activities are characteristically found among friends. And much of Part III of this volume will be devoted to exploring the ways that friends share activities, including conversations, via communication technologies, and the risks and rewards associated with sharing that occurs through these media.

Friendships are characteristically informal and unstructured, making it difficult to specify in advance what counts as an activity characteristic of friendship. The same holds true for features of friends. Anthropologists, for example, sometimes have difficulty identifying friendships (and even whether the institution exists in a given culture) because one's friends overlap with one's kin, colleagues, and neighbors, (Rich 1980; Uhl 1991; Paine 1969) and because the parameters of friendship are highly subjective and individual, ultimately determined by the friends themselves, leading one theorist to conclude,

> Inasmuch as friendship is recognized as a social relationship, it is an institution in the limited and rather loose sense of bestowal of recognition; and this is commonly the extent of its institutionalization in our culture, where it amounts to a kind of institutionalized non-institution.
>
> (Paine 1969, 514)

By reversing the direction of explanation, one can avoid trying to specify the unspecifiable. Rather than look for the activities that characterize friendship, we should conclude that friends valuing sharing activities (partly) explains their being friends. This frees us from the difficult task of attempting to identify necessary or sufficient activities for friendship.

Furthermore, sharing activities dovetails with the previously discussed feature of mutual responsiveness: valuing an activity in virtue of its being shared necessitates that one remains essentially open to the other's input, and that one's vision of the output be determined in part by the other's contributions. At the level of particular projects as well as long-term values, many of friends' goals and activities are shared and interdependent, and they have a shared and standing interest in sharing such joint pursuits. Since joint participation in a shared goal gives one some reason to think the participants are parts of a whole, and shared activity is characteristic of friendship, we have further reason to think friends are parts of a composite whole: a friendship.

Finally, consider mutual shaping of character and self-image. If my friend likes my dry sense of humor, I may start thinking of myself as someone who possesses this kind of wit, will be more likely to display it in future, and it will start to play a role in how I understand myself. Friends may also contribute to each other's growth and development, by highlighting and drawing out potential for change. Thus, friends not only respond to each other's interests and values, but also systematically influence each other's development, a phenomenon that has been explained in various ways (Cocking and Kennett 1998; Badhwar 2005).

Some recent literature on friendship and self-knowledge (Osborne 2009; Biss 2011) introduces additional reasons to think that friends are co-participants in learning and knowing, not merely as passive reflectors-back of traits we already possess, nor simply as foils against which to compare ourselves. Both positions are commonly ascribed to Aristotle; this may be an unfair and inaccurate attribution (Osborne 2009). Friendships also give us opportunities to exercise and display characteristics such as wit, by jointly engaging with friends in such paradigmatically friendly shared activities as joking around and bantering, giving us the opportunity to witness these features in ourselves while simultaneously developing and strengthening them (Biss 2011). The role friends play in shaping and reinforcing each other's character traits and self-image gives us an additional reason to think friends are parts which together compose friendships.

Mutual responsiveness and interdependence on a number of fronts, from values to activities, to one's own conception of oneself and one's development, have been identified as key components of friendship in a number of theories, and furthermore are plausibly important to lay conceptions of close friendship. Being a friend means some of the person's interests, activities, and the shape of his character are partially determined by his friends. Thus, taking friendships to be wholes offers distinctive explanatory advantages in understanding the behavior of the friends as parts by showing how these features are related to each other, as well as the intuition that friends share identity. Friends' sensitivity to each other, reciprocal mutual concern, and valuing of shared activities bind them together into a unified whole.

3. Friends or Enemies?

The parts/whole account of friendship offers several ways to differentiate friendship from other sorts of personal relationships, especially those that are intuitively unfriendly. In a certain sense, enemies may end up with coordinated interests. If Moriarty values something, that may be reason enough for Holmes to find it repugnant. But enemies will not often value shared activities in virtue of their being shared, nor will they share interests and projects with each other. In fact, their relationship is characteristically one in which their interests and projects are on collision courses with each other. Or consider the domineering spouse, whose subservient partner takes on new interests, but has to set aside his or her old ones, while the domineering partner remains inflexible. Furthermore, the subservient partner's self-image ends up being shaped by the other, but not the reverse (or at least not to the same degree). In both the dedicated-enemies and domineering-relationship cases, the "parts" of the relationship fail, in important ways, to function properly as parts. Thus, friendships stand out from these other, less friendly relationships, for exemplifying not just some but all of the characteristics mentioned, and to an especially high degree and across a wide range of circumstances in particularly close friendships. Furthermore, each of these features is not merely present in friendship, but paradigmatically valued by friends. Friends do not merely coordinate their values, but think it important that they do so. Friends do not merely share activities, but value this sharing. Friends do not merely change who they are in response to friends' observations, but do so because they find their friends' perspectives valuable.

Other sorts of cases may be less tractable to this analysis, but can, I think, ultimately be accommodated by the approach I advocate here, with the addition of a normative component. In practice, the line between friend and enemy can be blurrier than we like to acknowledge in theory, hence the existence of portmanteaus such as "frenemy". In such cases of apparently toxic friendship, perhaps the best thing to say is that although such people do constitute an organism, it is one which is unhealthy because belonging to it is bad for its component parts. This would make friendship not merely an ontological kind, but also a normative evaluation of the kind: friendships are the relationships that fit the model and are good for their members. But perhaps this is not such a strange thought: friendship overlaps with many other relationships, and to add that someone is, in addition to being a co-worker, classmate, or sibling, also a friend, adds a positive appraisal to the relationship described, as I have noted earlier. Insofar as the point of this project is to give an account of how people can consistently make friends out to share identity, and to differ, this is compatible with there being other situations wherein people's interactions inform their identities.

One might wonder whether identification of friends must be reciprocal: whether A can legitimately identify as a friend of B even though B does not so identify for A. My answer is that it must be reciprocal. I appeal here to the intuition that there is no such thing as unrequited friendship, even though there may be unrequited friendly feelings or desire to be friends. In some sense, this may seem like a stipulative definition of friendship, but it does important normative work, explaining friendship as a situation that requires the mutual flourishing of friends in order to be successful. It may be that early in the development of a future friendship, one person harbors more friendly feelings than the other; on my account, while that may play an important part in the story of how the friendship emerges, it does not exist until both friends reciprocate.

4. Many-One Identity: Parts and Wholes

We have yet to reconcile the apparent contradiction between "self" and "other". Turning to metaphysics, we find a parallel concern, the resolution of which suggests a solution for a theory of friendship. The apparently contradictory requirements that some things be both identical and differing have motivated the thesis known as Composition As Identity (CAI), a theory of identity between parts and whole that captures both their sameness and difference. Here, I use Donald Baxter's version of CAI. Baxter has argued that many apparent contradictions associated with some common theories of parts and wholes can be resolved by embracing a compositional account of identity (Baxter 1988, 2005).

Although CAI has historically been associated with other controversial metaphysical theses, such as Universal Mereological Composition (UMC) (which holds that any combination of items can be understood to compose a whole, such as David Lewis' infamous "trout-turkeys", composed of the front half of a trout and the back half of a turkey), it is distinct from them, and can be adopted without also requiring that one embrace UMC (Lewis 1991). UMC is a claim about what it takes to compose an object; CAI is a claim about the relationship between composed objects and their parts. One can have stricter standards for what it takes for parts to compose a whole and still endorse CAI. In fact, one can hold that composition conditions vary depending on the kind of composite object under consideration: perhaps there are very different relationships at work in water particles composing a cloud versus organs composing an organism. In the preceding portions of this chapter, I have sketched an account of conditions under which certain kinds of parts—organs—compose certain kinds of wholes: organisms. My purpose in introducing CAI is to make sense of the intuition that friends are unified even though they are different, by exploring how intuitions about unity and difference inform our more general theories about the relationship between parts and wholes. To the degree that this makes sense of intuitions of friendship, it may give advocates of CAI reason to explore its application in other kinds of wholes than the simple material items more commonly discussed in the literature.

Baxter argues that some paradoxes occur in standard accounts of identity (Baxter 1988). In cases of fusion, where two things become one thing, the product of the fusion seems to be identical with the things fused, but the things fused, which should be—on any reasonable account—identical with themselves, differ from each other prior to the fusion. Suppose one is making cookies, and squishes two lumps of dough together to make one cookie. If identity is transitive, the fused item—the cookie—seems to be identical with two different things (the lumps of dough), but by hypothesis the pre-fused things are not identical with each other.

Intuitively, one wants to say that in a case of fusion the two items each becomes a different *part* of the fused whole (perhaps they become different halves or regions of the cookie) while remaining identical with themselves. If parts are not identical with the whole, then it is consistent to say the two fused items become two different parts of the whole. But now the fusion seems to be a third thing, independent of its parts (Baxter 1988). A parallel problem holds for friendship: even if one takes friends to compose a friendship, friends seem to be different parts of the friendship, and so such a theoretical approach only preserves difference between friends at the expense of friendship's really unifying the friends. Perhaps one could say that friends really do join or merge into a whole, but only at the cost of rejecting differences between friends.

But this claim that parts are not identical with the whole is troubling. If the whole is something besides the parts, then we should be able to take the parts away and still have a whole. But if you take away the dough, there is no cookie left. And, whenever we have parts, we should in addition have something else: a whole. But, points out Baxter, we do not count like this (1988, 200–1). If one is dividing up a tract of land into parts, one does not sell off the portions but retain ownership of the whole tract: the tract is the parts (1988, 197). And if each can in a six-pack costs 50 cents, one does not pay $3 for the six cans, then another $3 for the six-pack (1988, 200). Although these "wholes" might not appear terribly robust, the same point holds even for more cohesive entities. One would not be expected to pay twice for a cow: once for the animal, once for its parts. The same holds true for friendship: it would be absurd to think that a friendship still exists even without the friends. To avoid this, we can suppose that the whole *is* the parts, and that many parts collectively are identical with a single whole. This would make many things (the parts) identical with one thing (the whole). This resolves some puzzles about parts and wholes, but requires a different theory of identity.

Baxter advocates that we take it as primitive that the same phenomenon can be both one unified thing, and many distinct things. Whereas what he terms the "received view" (Baxter 2005, 377) would hold that, for an object with two parts, there are really three things—the two parts plus the whole—on Baxter's account we have two parts and one whole, but these are not distinct and so do not add up to three different things.

Baxter introduces the technical concept of a "count" to flesh this out. The number of things we can count is always, he says, relative to a given standard for counting (Baxter 1988). Consider a six-pack. If a standard counts cans, there are six things there. The six-pack will not be counted, since it is not a can. If we count six-packs, there is just one. Each of the cans is not a six-pack and so it is not counted. Many of the more standard features of identity, including transitivity of identity, hold (only) within a count. But many-one identity consists of what Baxter calls "cross-count identity": identity across counts. The six cans are cross-count identical with the six-pack just when you can count the same region of the world and get six cans, or one six-pack.[1] Again, the point applies to other kinds of objects: there may be 206 bones in the human body, but if one asks "how many things are there?" of a skeleton, it would be odd to answer "207 . . . the bones plus the skeleton". The skeleton is the bones, even though the femur is nothing like the skull.

One might object, at this point, that we do not speak as though parts are identical with each other. No one thinks people's hearts are identical with their livers, or that a car's steering wheel is identical with its carburetor. But this problem is avoided by keeping track of "counts". When counting parts, the parts are distinguished from each other. A heart is a part, and so is not identical with a different part (such as a liver) on a part-count. But where both compose the same person, they are identical with that person, and there is just one person, so on a person-count both the liver and heart "count" as the same thing: the same person.

On my account of friendship, drawing on Baxter's theory of identity, friendships are composite objects with friends as parts. Where we count two people, and one friendship, the people counted, the friends, are cross-count identical with the friendship. There are two people there: you can still count two, using "person" as your counting standard. There are two friends there, as well, although their being friends is contingent on your being able to count a friendship as well. The same entity can be both a person and a friend.

Of course, a person can simultaneously be many other things as well: material object, protein-based life form, mammal, tax-payer, Republican, and so on. What the entity counts as depends on what is being counted. But we cannot add across counts to get three distinct objects, any more than we can count six cans, and then add the six-pack to get seven items.

This allows us to reconcile something's being both self and other without contradiction. Friends as parts turn out to be identical with the friendship they compose, and the friendship identical with them. Because each part is identical with the whole, the friends are identical with each other by way of being identical with the same whole, even though they are different parts of the whole.

5. On Identity and Identification of Friends

It is important to keep track of the relationship between identification and identity. To identify with something, psychologically, involves (very loosely speaking) a felt similarity to another that contributes to a tendency to empathize with them, to feel what (we imagine) they feel. When the hero or heroine of the film suffers a loss, we may identify with their plight. Of course, this is not sufficient to change *our* identities. However, to identify as something—such as a woman, or an academic, or an athlete—one can do either of the following: report something about oneself OR pick out a goal for oneself about what one will become. The disjunction is not intended to be exclusive: identification in this sense can involve both finding a potential and making it a goal to actualize this potential.

It is tempting to read identity in friendship as identifying with one's friend in the first sense, and doubtless plays a role in friendships (mutual responsiveness, as I defined it earlier, probably involves this sense of identification). But friendship also involves a different kind of identification: one identifies, or gains the capacity to identify, as a friend.

It is this second sense I am primarily interested in, both as a report and as a goal for people. What does it mean to identify as a friend as well as to identify with one's friend? My contention is that it involves more than merely identifying with one's friend, and that the identity invoked in unity accounts of friendship is more than mere felt similarity. My reason for thinking this is that we routinely identify with movie characters, celebrities, and characters in stories without taking ourselves to be friends with them. If friendship's identity were merely identifying with, then this would be peculiar, to say the least—it would make the unity account conflate friendship in the robust sense with a kind of passing felt similarity which we widely acknowledge to be transitory and context-sensitive. On the other hand, to postulate that friendship involves more than mere identification with another requires a plausible alternative account of what friendship is, and union accounts owe us this.

My theory is that to identify as a friend is to identify as a part of a composite object, a friendship, which (owing to an identity relation between parts and a whole they compose) amounts to a kind of literal sharing of identity with one's friend, in virtue of jointly composing said entity.

Claiming that two different friends can be the selfsame friendship has implications for our understanding of self-interest, because we will have to be careful to specify which sense of 'self' is being invoked in any given case. On my account, the friend is both the person and the friendship. So, the sense of 'self' invoked (person, or friendship) by considerations of 'self'-interest can vary. Acting in the interest of the friendship is acting in one's self-interest, because one is a friend and is thus part of the friendship, but acting in the interest of the person is also acting in one's self-interest, even when doing so is in tension with the continuation of the friendship. We might distinguish between broad and narrow selves, as (Baxter 2005) does, to make discussing this clearer. This account of self-interest helps to unpack what is involved with friends taking on each other's interests as background considerations, and the ways that one's interests can be advanced or hindered by a loved one's interests being promoted or frustrated, as Hampton indicates in the passage quoted at the beginning of this chapter. Friends use both a broad and narrow sense of self in evaluating their self-interest.

Unless we flesh out what is meant by 'self'-interest, it looks like taking friendship to involve adopting friends' interests and concerns will make conflict between friends, or between one's interests and one's friend's, inexplicable. Distinguishing between broad and narrow senses of self, meanwhile, lends itself naturally to the task.

It would be silly to insist that, because one takes one's friend's interests as background considerations or because one's interests are advanced or hindered by the success or failure of a friend's projects, that one's interests can never conflict with those of a friend's. But even one's own interests can conflict. On my account, a conflict between my interest and my friend's is relevantly similar to conflicts between my own interests, because conflicts within my own interests (as a person) are conflicts in narrow self interest, and conflicts between my interests and my friend's are conflicts in broad-self-interest. This also means that, just as I can sacrifice some of my interests for others without thereby sacrificing my interests across the board, I can sacrifice my interests for my friend's without thereby being said to sacrifice my self-interest in one sense (in the way that a soldier in a foxhole who throws himself on the grenade to save his friends, or the parents who forgo eating to feed their child, appear to promote their interests in one sense while sacrificing many of their interests). My account rules out is that sacrificing one's narrow self-interest for a friend could be *completely* selfless, but this is a point I am happy to concede as intuitively incompatible with a healthy friendship. I am persuaded by Jean Hampton's arguments in "Feminist Contractarianism", where she engages in a lucid discussion of some of the problems with self-abnegation and self-sacrifice in personal relationships (Hampton 1993a). These conclusions inform some of the discussion in Chapter 9.

Not every part of an organism needs to always work together, or work together successfully, toward some purpose or goal, in order to compose an organism. Even in human bodies, organs can work together in some ways but thwart each other in others. However, the well-being of the organism ultimately depends on both the individual well-being of its constituent parts, and their ability to work well together. Likewise, for friendship. Friendships thrive when friends both work well together and doing well—or as well as they can—together. The flourishing of the friendship, not just its identity conditions, explains why friendship seems to be a good thing for human beings.

This account will not, all on its own, fully explain why friends care about each other, because identity is not sufficient to give one an interest in something, especially not an all-things-considered interest. Finding out that you are a pathological liar is not in itself grounds for you to have an interest in lying; upon realizing this, one may then devote oneself to resisting this tendency. But it does explain how friends can see themselves as unified while *at the same time* recognizing, respecting, and valuing each other's differences. This is a factor that will become important in Part III, when discussing the nature and quality of connections via social media.

6. Richness and Variety as Sources of Value in Friendship

This metaphysical approach can help model much of what we think is valuable about friendship. Friends with complementary but differing character traits, different backgrounds or areas of expertise, and different interests and values can enrich our lives in ways that would be impossible if friendship thrived only in a monoculture, where each person was exactly like the other.

Friendship, on my account, involves each friend identifying as a part of a friendship, and thus as the friendship, and in virtue of the friendship, identifying with one's friend. This helps make sense of the way friends' interests are paradigmatically interrelated, coordinated, and interdependent. Each friend sees herself as both a part of the friendship and the friendship, and thus also as the friend. By seeing oneself as the whole which includes both friends, one sees oneself as now expanded or extended to include the friend. This eventually helps explain the importance to friendship of valuing and pursuing joint projects, including the interdependence of reasoning and openness to revision that implies. It also explains why we think friends can be an important and valued part of one's life.

If friendship is about expansion of self, and friendship is a valuable institution, then the value of this expansion needs explanation. Simply having *more* of oneself is one possible source of value, and consistent with Aristotle's original and ambiguous discussion in Book 9, Chapter 9 of the *Nicomachean Ethics*, where he spells out the benefits of expansion of self in terms of increased capacity for continuous activity (Aristotle 1999, 1170a). This may appear to assign friends a merely instrumentally rather than intrinsically valuable role in one's life, but it does not. Friends, by extending a person and becoming "other selves", are intrinsically valuable as components of an intrinsically valued entity. They are not mere means to an independent end.

This expansion, however, is impoverished if it is limited to duplication of a person's features. Bernard Williams argues that "differences of character" are of central importance in personal relations, and not obviously compatible with an Aristotelian account. "Once one agrees that a three-dimensional mirror would not represent the ideal of friendship", says Williams,

> one can begin to see both how some degree of difference can play an essential role, and, also, how a commitment or involvement with a particular other person might be one of the kinds of project which figured basically in a man's life . . . something which would be mysterious or even sinister in an Aristotelian account".
>
> (Williams 1981, 15–16)

But if we incorporate differences between friends into a theory of friendship, this makes the expanded entity more complex, not just bigger, defusing Williams' objection. In friendship, one takes on a friend's different talents, interests, and concerns without sacrificing one's own. Friends become identical with the friendship by identifying as both parts of, and as, the whole. They are now more than they were, before they became parts. This means that friends together are not just capable of more continuous activity but can lend differing perspectives and talents to a shared project. They can offer complementary traits, as, for example, where one friend's courage is supplemented by another friend's carefulness. These features are not available simply by scaling up one person's efforts. Shared activity, including a shared life, is thus enriched by at least some differences between friends. The ability to share an activity despite differences between individuals' situations and dispositions will become relevant when discussing social media, in Part III.

On a simple unifying model of friendship, it is difficult to capture the enriching value of differences between friends. But a simplistic model's inability to capture this richness is not good grounds for concluding that identity between friends *cannot* be shared. The ability to accommodate and explain such richness is one of the significant advantages offered by a part/whole identity theory of friendship. Insofar as we aim to articulate the ideals of friendship, it is important to capture the intuition that it is admirable for people with great differences to be friends with each other. That this theory allows us to do so without coming into conflict with intuitions about shared identity as another ideal of friendship counts in its favor.

One might object that the role of difference in friendship is overstated, because we tend to magnify small differences when assessing our own relationships, and overlook the deep similarities that hold between ourselves and our friends.[2] But the project here is not merely to describe what actually holds in ordinary friendships, but to articulate what we aspire to—what constitutes an ideal friendship. Here, it is clear that friends who differ in nationality, background, politics, and interests are frequently upheld as exemplars of excellent friendship, and so we ought not to overlook this as a feature of our ideals.

7. Conclusion

I began by noting apparent tensions between self and other in friendship, exemplified by Aristotle's claim that friends are "other selves". Both are highly desirable in a theory of friendship, and yet it was difficult to reconcile the two. I considered several reasons to think that friendships are genuine unities, then appealed to a metaphysical account of parts and wholes to show how it is possible to have friends be both "self" and "other", by being different parts which are nonetheless identical with the whole they compose, the friendship. The friendship is identical with the friends, not something other than them, and the friends collectively are identical with the friendship. We saw how—surprisingly—such an account makes for a natural explanation of the way that variety and difference among friends can contribute to a friendship. Friends contribute different strengths and perspectives to enrich the friendship, and by becoming part of the friendship, one makes such richness one's own.

It is unlikely that a liar and an honest person could have a successful friendship, and their differences appear to be relevant to explaining such a friendship's failure. But a friendship between a cautious person and a more adventurous individual does not necessarily face the same kind of disadvantage, so the problem with the first case cannot merely be difference. I conclude that each friend, as a part of the friendship, fulfills different functions, each contributing to the flourishing of the composite object. Undesirable differences may be undesirable because they prevent friends from working well together as parts of a coherent whole.

Composite objects such as friendships complicate our social ontology, leading many to focus on individual concerns. But resources for thinking about these complex entities can be found in contemporary work in metaphysics. The explanatory advantages of adopting this metaphysics of friendship justify its inclusion in our ontology.

What does this mean for ethics? After all, valuing friendship means valuing people's interdependence of interests and values, and their mutual influence on each other. This gives us reason to take the influence of social media seriously, and will inform much of what takes place in Part III. Some people are, so to speak, a good influence on their friends, but others less so. In the following chapter, I argue that despite recent objections that friendship can be hazardous to morality based on exactly this sort of interdependence, we have reason to think that ideals of friendship commit us to morality.

Notes

1 Note: this is not the same thing as making counts relative to sortals, but in some cases a count standard can be a sortal.
2 I thank Santiago Mejia for raising this issue.

Bibliography

Aristotle. 1999. *Nicomachean Ethics*. 2nd ed. Translated by Terence Irwin. Indianapolis, IN: Hackett Publishing Company. https://books.google.com/books?id=-Mf5XV8q6CgC&dq.

Augustine. 1993. *Confessions*. Translated by Frank Sheed. Indianapolis, IN: Hackett Publishing Company. https://books.google.com/books?id=20GvAwAAQBAJ.

Badhwar, Neera Kapur. 2005. "Love." In *The Oxford Handbook of Practical Ethics*, edited by Hugh LaFollette, 42–69. Oxford: Oxford University Press. https://books.google.com/books?id=emstfpfaT2wC.

Baxter, Donald L.M. 1988. "Many-One Identity." *Philosophical Papers* 17 (3): 193–216. doi:10.1080/05568648809506300.

———. 2005. "Altruism, Grief, and Identity." *Philosophy and Phenomenological Research* 70 (2): 371–83. doi:10.1111/j.1933–1592.2005.tb00532.x.

Biss, Mavis. 2011. "Aristotle on Friendship and Self-Knowledge: The Friend beyond the Mirror." *History of Philosophy Quarterly* 28 (2): 125–40.

Cocking, Dean, and Jeanette Kennett. 1998. "Friendship and the Self." *Ethics* 108 (3): 502–27. doi:10.1086/233824.

Cooper, John M. 1977. "Aristotle on the Forms of Friendship." *The Review of Metaphysics*, 619–48. doi:www.jstor.org/stable/20126987.

Friedman, Marilyn. 1998. "Romantic Love and Personal Autonomy." *Midwest Studies In Philosophy* 22 (1): 162–81. doi:10.1111/j.1475–4975.1998.tb00336.x.

Gilbert, Margaret. 1990. "Fusion: Sketch of a 'Contractual' Model." In *Perspectives on the Family*, edited by Robert C. L. Moffat, Joseph Grčić, and Michael D. Bayles. E. Mellen Press. https://books.google.com/books?id=zAbxAAAAMAAJ.

Hampton, Jean. 1993a. "Feminist Contractarianism." In *A Mind of One's Own: Feminist Essays on Reason and Objectivity*, edited by Louise M. Antony and Charlotte Witt, 104:337–69. Westview Press. https://books.google.com/books?id=eU6F8N4YbiAC.

———. 1993b. "Selflessness and the Loss of Self." *Social Philosophy and Policy* 10 (1): 135–65. doi:10.1017/S0265052500004052.

Helm, Bennett. W. 2010. *Love, Friendship, and the Self: Intimacy, Identification, and the Social Nature of Persons*. Oxford: Oxford University Press. https://books.google.com/books?id=8s0UDAAAQBAJ.

Lewis, David K. 1991. *Parts of Classes*. Cambridge: B. Blackwell. https://books.google.com/books?id=mBeFQgAACAAJ.

Osborne, Catherine. 2009. "Selves and Other Selves in Aristotle's Eudemian Ethics Vii 12." *Ancient Philosophy* 29 (2): 349–71. doi:10.5840/ancientphil200929230.

Paine, Robert. 1969. "In Search of Friendship: An Exploratory Analysis in 'Middle-Class' Culture." *Man* 4 (4): 505–24. doi:10.2307/2798192.

Rich, George W. 1980. "Kinship and Friendship in Iceland." *Ethnology* 19 (4): 475–93. doi:10.2307/3773153.

Schoeman, Ferdinand. 1985. "Aristotle on the Good of Friendship." *Australasian Journal of Philosophy* 63 (3): 269–82. doi:10.1080/00048408512341881.

Sherman, Nancy. 1987. "Aristotle on Friendship and the Shared Life." *Philosophy and Phenomenological Research* 47 (4): 589–613. doi:10.2307/2107230.

———. 1993. "The Virtues of Common Pursuit." *Philosophy and Phenomenological Research* 53 (2): 277–99. doi:10.2307/2107769.

Stocker, Michael. 1981. "Values and Purposes: The Limits of Teleology and the Ends of Friendship." *The Journal of Philosophy* 78 (12): 747–65. doi:10.2307/2026245.

Telfer, Elizabeth. 1970. "Friendship." *Proceedings of the Aristotelian Society* 71: 223–41. doi:10.1093/aristotelian/71.1.223.

Uhl, Sarah. 1991. "Forbidden Friends: Cultural Veils of Female Friendship in Andalusia." *American Ethnologist* 18 (1): 90–105. doi:10.1525/ae.1991.18.1.02a00040.

White, Richard J. 2001. *Love's Philosophy*. Rowman & Littlefield Publishers. https://books.google.com/books?id=iQNvAAAAQBAJ.

Whiting, Jennifer E. 1991. "Impersonal Friends." *The Monist* 74 (1): 3–29. doi:10.2307/27903221.

Williams, Bernard. 1981. "Persons, Character, and Morality." In *Moral Luck: Philosophical Papers 1973–1980*, edited by James Rachels, 1–19. Cambridge Paperback Library. Cambridge, UK: Cambridge University Press. https://books.google.com/books?id=wMGW2Ehldp8C.

3 Why Bad People Can't Be Good Friends

1. Introduction

In the preceding chapters, I have made the case that friends are best conceived of as parts of complex social entities, friendships, bound together by interdependence of values, emotions, and mutual influence. I have done so in the service of providing ethical guidance on social technologies, especially those involving friendship. At this point, I face a rather serious objection. This very interdependence and mutual influencing of values, one might think, makes friendship a moral hazard. It can present us with reason to behave more badly than we would otherwise, and to develop our characters in worse ways than we would without our friends to influence us. As the saying goes, a friend helps you move, but a *good* friend helps you move a body. If that is true, then technologies and practices that facilitate friendship would end up being those that expose us to obstacles to moral development, and thus ethically problematic simply in virtue of supporting friendship. This objection can be overcome, or at least blunted substantially, if we can find reason to think that ideals of friendship commit us to the idea that the best friends are also morally good people. While particular non-ideal friendships could still present a temptation to act badly and corrode one's values, friendship itself would offer a reason to resist temptation and become the best person one can be.

But must the best friends necessarily be good people? On the one hand, as Aristotle puts it, "people think that the same people are good and also friends" (Aristotle 1999, 119, 1155a). On the other hand, friendship sometimes seems to require that one behave badly. For example, a normally honest person might lie to corroborate a friend's story. This seems to follow from much of what has been said about the importance of mutual influence and interdependence. What I will call here closeness, which I take to include sensitivity to friends' subjective values and concerns as well as an inclination to take their subjective interests as reasons for action, is characteristic of friendship. But this seems to require that good friends should be morally flexible, more so than is compatible with a virtuous character. This would imply tension between ideals of friendship and ideals of character. But there is an important connection between virtue and friendship which arises precisely from friends' closeness, when concern for well-being, another important feature of friendship, is also taken into account. This helps mitigate the tension and shows how friendship and virtue are interconnected. The connection in turn provides friendship-based reason to think the best friends must be good people, even though concerns of friendship may occasionally clash with other moral concerns.

Although the importance of character to friendship is sometimes defended by arguing that virtuous people tend to treat each other better, and/or that vicious people are likely to mistreat each other in ways incompatible with friendship (for example, to cheat, steal from, and lie to each other), on its own this seems insufficient for a robust defense of the role of virtue in friendship. Friends may need to treat each other well, but this is seems to be compatible with their being otherwise dishonest, cruel, thoughtless, and so forth, to other people. In such cases, then, virtue seems like merely one possible route to a good friendship, and not itself necessary for the friendship to be good. In what follows, I will focus on another line of argument, one which does not rely on empirical questions about whether people who are generally bad to others are capable of making exceptions when it comes to their friends.

Some accounts of friendship portray friendship as opposed or at least unconnected to morality (Cocking and Kennett 2000; Nehamas 2010; White 1999; and the modern version of friendship sketched in O'Connor 1990). Such accounts assume that friendship is about closeness, openness to each other's perspectives, and friends' coordination of values and activities. On these accounts, friendship is a distinct but non-moral good, whose norms can conflict with those of morality. But I think that this sells friendship short. Closeness is partly constitutive of friendship, but only partly, and while closeness on its own may pose a threat to a person's moral motivations, other features of friendship, including concern for friends' well-being and desire to promote their good, are also partly constitutive of friendship. These other features on their own, however, are not sufficient for good friendship without closeness, because in the absence of closeness, paternalism in excess of the ideal of friendship would be licensed. Both closeness and concern for well-being, then, are necessary for the best friendships, and closeness and concern for well-being together generate a connection between the character of friends and the quality of their friendship. Friends may be very close without expressing proper concern for well-being, or show concern for well-being without being very close, but neither arrangement represents an ideal of friendship. Those who take friendship to be independent of considerations about moral character thus sell friendship short as a good, by portraying it as being primarily about closeness, and miss an important connection between ideals of friendship and ideals of character. Even if one does not take friendship to be a moral good properly speaking (in the way that one might take fairness, for example, to be a moral good), one should still believe that friendship is closely intertwined with virtuous character; that good friendship requires good (though maybe not perfect) character.

Friendship, I have argued, is like an organism: its parts (the friends) must be closely aligned, interresponsive, and coordinated with respect to some important interests and values. But this closeness, though necessary for friendship, is not sufficient for its well-being, which is partly dependent upon the well-being of its parts. An unwell part (or vicious friend) necessarily means an unwell organism. Being friends with a vicious person results in an unhealthy friendship, one in which the closeness of friends negatively affects the friends' well-being and thus fails to fulfill the ideals of friendship in at least one important way.

2. Virtue and Good Friendship as Necessarily Connected

Aristotle's account of friendship entails a connection between virtuous character and friendship (and which most later accounts of friendship draw on for inspiration or opposition). As we saw, Aristotle argues that there are three kinds of friendship: friendships based on mutual usefulness (which he calls friendships of utility), friendships based on pleasure taken in each other's company (friendships of pleasure), and friendships in which friends value each other as good in themselves, which he calls friendships of virtue. These last, he says, "are likely to be rare, since such [virtuous] people are few" (Aristotle 1999, 123, 1156b). If virtuous people are relatively rare, then combinations of virtuous people are limited by the number of potential virtuous friends, and will, correspondingly, also be rare. Nonetheless, despite their scarcity, friendships of virtue are the best and fullest form of friendship (Aristotle 1999, 123, 1157a).

Perfect friendships then require fully virtuous participants, who seem vanishingly rare in real life. But John Cooper argues that a more moderate version of the idea is available: virtue friendships are characterized by the goodness of the friendship's being predicated on the (moral) goodness of its participants. Middling good people have middling good friendships, mediocre people have mediocre friendships, and very good people have very good friendships. In the limit case, morally perfect people will have perfect friendships (Cooper 1977). So the theory may be read as a description of an ideal of friendship, which people can more or less closely approximate depending on the state of their own and their friends' characters. One need not read this as implying a strict correlation between the goodness of friends' characters and the goodness of their friendship. Some virtues and vices may have a more direct bearing or a more significant impact on a friendship than others. But the Aristotelian can and should say that thoroughly bad people cannot be good friends, while people with better characters will tend to have better friendships, all things being equal.

It is a basic assumption of most eudaimonist theories of ethics, including Aristotle's, that character matters to one's well-being: virtue is important for the best life, and vicious people do not live the best lives. If one cannot show that an apparent virtue is related to the good life of the individual who possesses it, this is generally taken to cast doubt on its status as a virtue. Philippa Foot, for example, famously said that "if justice is not a good to the just man, moralists who recommend it as a virtue are perpetrating a fraud" (Foot 2002, 126). This premise then yields at least one plausible connection between good friendship and virtue. Good, virtuous people have the best shot at living well, being equipped to value and promote their own and other people's good, and better people are better at this, all other things being equal. Friendship involves concern for friends' well-being, and so it is, then, no accident that good friends are called *virtue* friends in Aristotle's theory, and that the best people make the best friends.

3. Against Highly Moralized Friendship

We might worry, however, that tying friendship too closely to virtue rules out an intuitively important feature of a good friend: your responsiveness to their interests, even when these interests seem to be in conflict with your morals. Dean Cocking and Jeanette Kennett (Cocking and Kennett 2000) charge that one important feature of friendship is being directly receptive and responsive to friends' concerns in a way that precludes what they characterize as *filtering*: evaluating the friend's concerns in terms of one's own conception of morality, before taking them as reasons for action. While friendship is a valuable human relationship, they conclude, its norms and ideals clash with those of morality: in fact, it necessarily involves moral danger.

They appeal to a bit of dialogue from Jane Austen's *Pride and Prejudice* to illustrate the point. Elizabeth Bennett tells Mr. Darcy that "A regard for the requester would often make one yield readily to a request, without waiting for arguments to reason one into it" (Cocking and Kennett 2000, 285). Similarly, Michael Stocker argues that in becoming a friend, one becomes the kind of person for whom the friend's concerns are taken directly as reasons for action (Stocker 1981). Though these reasons of friendship may not always override competing considerations, they nonetheless characteristically figure in friends' deliberations and choices. In fact, failure to be open to the other's point of view seems inherently unfriendly. "If I always insist upon my own point of view, or really refuse to listen, then I have not made a serious commitment to the other person in the strong sense that friendship requires", argues Richard White (1999, 82). But listening to morally bad reasons seems incompatible with excellent character. Note that the concern I am raising here is not merely one of epistemic modesty (about the importance of being open to other viewpoints, say, or willingness to concede that one might be wrong), but rather that friendship involves a commitment to take a particular other person's reasons as prima facie valuable in their own right, in some special and strong sense above and beyond general norms of discourse and interpersonal conduct. But this can be hazardous to one's character. For an ordinary example of this, consider that being a good friend to a person with a particularly sharp wit may involve laughing at his jokes, reinforcing this trait in the friend while also leading one to share, somewhat indirectly, in his cruel humor.

This kind of closeness then seems incompatible with being a fully virtuous individual. Friendship requires relinquishing some control over the shape of one's life, desires, and goals. For contrast, Cocking and Kennett invite us to imagine someone whose responsiveness to you is "subordinate to and filtered through moral considerations" (Cocking and Kennett 2000, 295). Such a person seems to be too cold and inflexible, too much of a goody-two-shoes, to make a very good friend, and it is her goodness that gets in the way of her friendliness. For her to display the appropriate warmth and responsiveness that characterize good friendship, she would have to set aside her moral "filter" when it comes to the friend's concerns. But that would leave her susceptible to the friend's vices.

If closeness involves taking the other's perspective and subjective interests as prima facie valuable, even where one would ordinarily be left cold by these considerations, one's own values and judgments about what is right may change over time, and such change may not be an improvement: as Alexander Nehamas puts it, rather starkly, "I may never realize that as a result of our relationship my judgment was gradually debased and that I may find myself happy to have become someone I would have hated to be had I not submitted to you" (Nehamas 2010, 280).

To avoid this fate, one may choose morally good people as friends, so as to limit the likely number of practical conflicts between morality and friendship. But that only shows why good people ought not to choose bad people as friends, not why bad people make bad friends, at least to people who have no special commitment to morality and/or who share the same vices. It also shows that morally, one ought to choose good people as friends, but not why bad people make bad friends as such. To reach this stronger conclusion, we would need to see why the best possible friendships are necessarily between virtuous individuals, as Aristotle claims. But this claim seems implausible without further defense.

Virtue, for the Aristotelian eudaimonist, is an excellence of the best possible human life, and so desiring virtue for someone means desiring what is best for them. To defend the idea that norms of friendship entail that good people and good friends are the same people, one must persuasively show why virtue is a necessary component of good friendships rather than accept that a perfectly good friendship can morally corrupt a person, reinforce one's own vices through shared celebration of flaws, or simply be unrelated to the moral character of friends. Closeness alone seems to offer no such reason.

4. Why Mutual Concern for Well-being Is Inadequate for Friendship

Closeness alone, as I have indicated, seems insufficient for a good friendship. Friendship includes not merely responsiveness to a friend's subjective interests, but also concern about friends' well-being. It is by no means guaranteed that responsiveness to a friend's subjective concerns will coincide with concern for the friend, especially if the friend's subjective values fail to adequately track what is good for them. As Diane Jeske puts it,

> if Henry is not good, it seems that I cannot promote his ends as "independent goods"—his ends simply are not goods at all. This sort of objection, however, confuses concern for a friend with concern for his subjective ends. I can care about Henry, be concerned to promote his well-being, and yet recognize that he has chosen to pursue harmful or trivial ends. My concern for him will lead me to try to help him to revise his ends—my concern for him is a concern that he pursue worthwhile ends, or, in other words, that his subjective ends correspond to his objective ends. Concern for a person need not involve valuing the ends she has chosen for herself; parental concern for children and especially for teen-agers is a good example of concern for a person coming apart from concern for her chosen ends.
>
> (Jeske 1997, 65)

Jeske ultimately concludes that concern for a person's objective ends may not always keep friendship consistent with morality. But others take the spirit of her concern to motivate an account of friendship that traces a necessary connection between virtue and concern for well-being characteristic of friends. In this vein David O. Brink argues concern for a person provides a natural account of the relationship between friendship and virtue (Brink 1999). Just as people concerned about their own well-being will have reason to become virtuous, on a eudaimonist account, they are concerned for each other's well-being and so seek to promote virtue in their friends. In his account, mutual concern yields a natural connection between friendship and virtue, but it does so only at a cost that I think is too high.

Brink argues that it would be a mistake to think that friendship of virtue requires one to love people just for the virtue they (already) possess, because this would seem to make friends valuable just as *bearers* of virtue: if that were true, then what one really loved would not be the friend, but the virtue that friend exemplifies, a concern he gets from Gregory Vlastos (Vlastos 1973). We saw some of the problems with this approach in Chapter 1, discussing Whiting's arguments for this kind of abstraction in friendship. Instead, Brink argues, what is most central to virtue friendship is that virtue friends care about each other. Given that they care for friends as they care for themselves, they want what is best for them. When we care about friends, argues Brink, we desire virtue for them, because virtue is importantly related to a person's well-being. Thus, while we need not have friends who are *presently* as virtuous as possible, argues Brink, "we can understand Aristotle's claim that interpersonal virtue-friendship reflects the comparative worth of friends as the claim that friends who care about each other for the other's own sake will prize and seek to promote the other's virtue" (Brink 1999, 273). He concludes that virtue friends, the best kind of friends, are characterized by virtuous manner of caring, rather than the sorts of people who enjoy such relationships.

But this seems unsatisfactory without taking into account the closeness discussed in the previous section. As Cocking and Kennett argue, in ordinary cases, it seems inaccurate to describe friendship exclusively in terms of concern for the friend's welfare.

> [T]he nature of the interest I have in her as my friend, and the reasons I have to act as her friend which I do not share with others, such as, for example, my colleague, is more naturally explained by my special receptivity to her direction than by any distinctively moral concern for her welfare.
>
> (Cocking and Kennett 2000, 287)

Brink's account, then, seems lacking because it avoids any "special receptivity to direction" that might come apart from concern for the friend's moral character. Chapter 1 gestured to some concerns with this approach. In this context, more associated problems become clear.

On Brink's account, it seems to be entirely possible that a good friend may, for instance, report a friend's tardiness to the boss, if doing so will be conducive to the development of the friend's character. For example, facing up to the consequences of his actions may give the friend a chance to reflect and repent, to reconsider life choices, and thereby become a better person. This sounds counter-intuitive as an account of a friendly action, however. Brink would either have to argue that it is *never* in a person's interest to be forced to own up to the consequences of his actions, or else that turning a friend in to an authority figure is in fact a friendly action, counter-intuitive as it sounds.

But it is troubling to suppose that because a person knows what will be good for his friend, their friendship entitles him to act against that person's wishes, for her own good. Such paternalism, while eminently appropriate in, for example, relationships between parent and child, or even (in limited circumstances) between teacher and student or therapist and client, seems to be incompatible with good friendship in a great many cases. There may be times when friendship allows or even calls for *some* paternalism, but there are important limits. It will be problematic if promoting a friend's virtue seems to require or allow for more paternalism than is plausibly characteristic of good friendship. A good theory of friendship will allow one to distinguish, then, between friendly (meaning: compatible with being a good friend) and unfriendly paternalism.

It seems important to respect and respond to a friend's subjective interests. In some cases, as a friend, you ought to support her even though you disagree. A good friend may date or marry a person you find unappealing, and yet friendship may require you to muster up some congeniality toward the partner, or at least to keep quiet about your distaste. Or, a friend might decide to pursue a career that you disapprove of, but your concern for your friend might compel you to support her, nonetheless. Friends may disagree, sometimes quite deeply, about politics, religion, and social issues, and yet, out of respect for their friends' perspectives, tone down or avoid voicing their objections in their friends' presence, or where their friends might be affected. (There are limits, of course, to how much one might be expected to tolerate, but some tolerance is, it would seem, desirable.)

Some tensions between sensitivity to a friend's subjective interests and what you take to be her well-being will turn out to be illusory, and in some cases paternalism will be justified even between friends: if what your friend is pursuing is dangerous, as a friend you may be obliged to say that you think this is a bad idea, present her with reasons to reconsider her plans, or even, in extreme cases, actively oppose her (suppose I hide her car keys when she's been drinking). This might seem sufficient to justify Brink's version of virtue friendship, and Jeske's example of parents and teenage children trades on the intuition that setting aside a loved one's subjective interests is sometimes justified by concern for their objective well-being.

However, some important constraints must be placed on such concerns to prevent counter-intuitive conclusions about the role of paternalism in friendship. After all, relationships between parents and teenage children are not paradigm cases of good friendship. Something like Brink's approach runs the risk of widespread paternalism, because when one acts purely out of concern for the friend's objective well-being, the friend is not consulted, nor is their perspective factored in. Thus, it seems lacking as a description of good friendship if it cannot plausibly explain how friendship may at least occasionally involve setting aside one's values (moral or otherwise) to support a friend, nor why closeness seems to require taking friends' reasons to be at least prima facie reasons for oneself. We could stipulate that what we are sensitive to is the friend's good as the friend himself conceives it. However, once we do this, we seem to be back at the problem Cocking and Kennett pointed out. Taking a friend's concerns as directly important to oneself, without an intermediary moral filter, leaves one susceptible to acting immorally or at least less morally than one might otherwise be inclined to do.

5. Valuing Friends Versus Valuing Virtue

If a theory lacks the ability to distinguish between promoting a friend's well-being, and promoting a friend's subjective interests (or at least respecting and supporting them), either too much paternalism seems to be endorsed, or friendship seems to be merely a matter of closeness. We need a finer-grained approach in order to accommodate both the intuitive point that friends are responsive to each other's interests, and the thought that friends are concerned for each other's objective well-being.

There are additional reasons to think friendship cannot and should not be contingent on a friend's virtue, presented in Vlastos's objection to Aristotelian virtue friendship, which was part of Brink's motivation for developing his own version of virtue friendship. Vlastos argues that in Aristotle's account, virtue friends are valued as bearers of virtue and are thus not themselves intrinsically valued, and this sounds quite unfriendly (Vlastos 1973). We can respond to Vlastos's challenge, however, by carving out a new role for virtue that makes the best friends (evaluated by the standards of friendship) neither (mere) bearers of virtue, nor recipients of paternalistic beneficence, nor yet major sources of moral danger, as the accounts surveyed so far seem to do. Furthermore, such an account can draw on reasons and concerns internal to friendship to explain the importance of character.

It seems intuitive that in valuing a friend, one values the person, not merely their virtue. And it seems natural to think that in valuing a friend as a person, one should take a friend's subjective values and concerns to be important and valuable; this explains the closeness emphasized by Nehamas, White, O'Connor, and Cocking and Kennett. This does not, however, rule out virtue as a concern for friends, because in valuing the person one does not (as Jeske points out) *merely* value that person's subjective perspective on the world; one also wants the friend to do well, and have a good life. And recall, for the eudaimonist, that the virtues are associated with the best life.

Grant that a friend's perspective can differ from one's own. Where subjective values are shared, clashes between friends' values and interests are minimized, and closeness is achieved. While we can imagine friends who "agree to disagree" about, for example, religious matters or social issues, this seems most successful when they agree about other central matters of value: integrity, honesty, and so forth. In the first chapter, we encountered the idea of minimal structure as a characteristic of friendship. Laurence Thomas points out that friendship is a "minimally structured" relationship, but

> minimally structured interaction will be harmonious only if the parties involved are sufficiently attuned to the way in which each other views and interacts with the world. . . . Successful minimally structured interaction requires a shared conception of the good.
>
> (Thomas 1987, 220)

The friend whose idea of caretaking is to leave a person alone will make a poor friend for the person who expects and offers more involvement in times of stress and hardship (and vice versa: the person who wants time alone to process things may feel smothered by attention from a friend who can't see and respect this). This, then, gives a reason internal to friendship to think the best friendships are those between friends who share at least some important values. Shared values promote closeness, which is central to friendship.

Values, however, influence how a person lives her life, and we want our friends to have good lives. So at this point, a eudaimonist should ask which values a person ought to have. Given that friends care about each other's good *and* are sensitive to each other's conception of the good, the best friendships to be in will be those in which the friends are both in agreement about what is good, and correct about what is good. Such friends will tend to promote each other's actual well-being, and not inadvertently encourage friends to pursue harmful or self-destructive goals.

For friends who are close but not virtuous, however, prospects are less sanguine. A gambler, for example, may promote his own pleasure at the expense of his bank account, and friends gambling together may jointly promote their pleasure at the expense of their finances. Even where direct sensitivity fails to adversely affect another's character, friends with vices fail to consistently promote their own interests, and thereby harm the interests of those who care about them, because in harming themselves they harm something of value to their friends. It seems incorrect to call people like the gambling partners above *good* friends, though they are directly sensitive to each other's interests, because their shared conception of the good ends up being bad for them, and so they inadvertently help each other to hurt themselves.

Although this point might seem tendentious, it is consistent with the majority of popular eudaimonist theories of virtue. Consider the possible ways that virtues can contribute to living a good life, and the ways that vice can prevent this. Some of the most common strategies are as follows. One may hold, as the Stoics did, that virtue is both necessary and sufficient for a good life. Or, following Aristotle, one may think that virtue is necessary but not sufficient for a good life; one cannot live a good life (or at least, not the best possible life) without virtue, but some "external goods" are also required; contingencies like health, resources, and social circumstance. Finally, one might hold that virtues are the most reliable strategy for living a good life, though one might end up with a good life by being lucky or pursuing less reliable or more risky means (or, alternatively, one could take the most reliable approach and fail anyhow). Combinations of these positions are possible (some virtues might be necessary for a good life because they are partly constitutive of it, others could be merely reliable means of achieving an independently specifiable good), but for our purposes it will be enough to take each in turn and see how direct sensitivity to a vicious person ends up exposing a person to harm, or potential for harm, in a way that is incompatible with the best sort of friendship.

Suppose the Stoic approach is correct, and virtue is both necessary and sufficient for happiness. Then, if one is close to a vicious person, and hence directly sensitive to their subjective concerns, one ends up less virtuous than one would otherwise be, and so less happy. The vicious friend thus makes one less happy, and this seems intuitively unfriendly. Or, suppose the Aristotelian approach is correct: virtue is necessary but not sufficient for happiness. Closeness to the wrong people can thus deprive a person of at least some virtues necessary for a good life, and this also seems unfriendly. Finally, suppose virtue is the most reliable strategy for living a good life. Vicious friends then encourage one (not deliberately, but simply by presenting one with their subjective concerns and values) to adopt strategies less reliably connected a good life; again, this seems unfriendly, and unfriendly in virtue of failing to promote friends' well-being, and in fact making them worse off.

But the story cannot end there, because the cases raised so far presume that having a friend *more* vicious than oneself is damaging, and hence unfriendly. One might wonder about cases where one's friend is only equally vicious. Consider the two gamblers, who in fact bond over and share in a vice together. Why think that this is incompatible with the best friendship, or at least, the best friendship available to them? Here, to respond, it is important to note that friends not only change us, but can reinforce traits in us, as well, by responding to us and by sharing our concerns. This is a good thing if these traits are those one ought to have (the friends who reinforce each other's thoughtfulness or bravery seem to be good for each other), but bad if you would be better off without those traits. Again, a quick survey of the possible connections between virtue and a good life shows that in each of the most popular versions of virtue ethics, one is worse off for having bad traits reinforced by a friend who shares those traits. If virtue is necessary and sufficient for a good life, vicious friends reinforce each other's vices and each makes it less likely that the other will move away from vice and develop the traits that would constitute a good life for them. That seems unfriendly. The same is true even if virtue is necessary but insufficient for happiness: by reinforcing traits that stand between people and the best possible life for them, one seems to harm them. Finally, even if virtues are merely the best possible strategy for living a good life, vices which are reinforced will make it less likely that one will live a good life, not a fate one ought to wish for a friend.

While vicious friends may display the characteristic sensitivity to each other's needs that we find in friendship, there is no reason to think that these cases are especially exemplary or choice worthy cases of friendship. In fact, there is reason to think that such friendships are flawed, because such friends cannot reliably or consistently promote each other's good, and in fact such friends tend to harm each other. This clashes with the common sense view that a good friend will not help you hurt yourself. Thus, at least major vices, the sorts of character flaws which are significantly detrimental to one's ability to live a good life, will clash with the requirement that friends show concern for well-being.

Finally, one smaller point about the interactions of friendship, virtue, and concern for well-being: just so long as a vice is harmful to the person possessing it, the vicious friend ends up harming a portion of her friend's interests by harming herself. So even in the event that friends' closeness is insufficient to influence each other for the worse, the friend of the vicious person ends up worse off simply because the person whose well-being they value is incapable of being as well off as they would be if they had better character. Although none of us are perfect, friendships are in general made better by virtue, and worse off by vice. Though some minor flaws and failings may have little impact on the quality of a friendship (someone who is a little vain may still be quite a good friend), major vices will be incompatible with good friendship.

6. Connecting Virtue and Friendship

One need not value a friend as a mere bearer of virtue, but friends should be virtuous to enjoy the best friendship. Friends can be "partners in crime", but these will not be the best sort of friendships because they will not consist in both friends consistently and wholeheartedly wishing and promoting what is good for each other, because their reasons for action are not consistently good even for themselves. This is not just a good self-interested reason to be selective in one's choice of friends; it is a reason internal to friendship to think character matters.

Direct sensitivity to another's concern does not rule out the possibility that bad people may be interconnected by sharing concerns. But it does give reason, based in concern for well-being (both one's own and one's friend) *not* to become close to another unless both parties are virtuous. As Aristotle predicted but Brink conceded, the best friendships will be limited by the number of virtuous people who can participate in them. This does not eliminate moral danger entirely, but it does provide a necessary connection between good friendships and good people; between standards of friendship and standards of character.

7. Conclusion

We might worry that accounts like Aristotle's overlook moral danger inherent in friendship; friendships are characterized by closeness and sensitivity to each other's needs and concerns, which may compromise one's commitment to living a virtuous life. But as Brink notes, friends are concerned for each other's well-being and so (on a eudaimonist account) have reason to promote each other's virtue. But without closeness, which entails sensitivity to a friend's subjective values and concerns, attempts to promote virtue can risk unfriendly paternalism. In friendship, it is important to consider both a friend's actual and perceived good. I conclude that direct sensitivity and objective concern for the friend's well-being together provide reason to think that bad people cannot participate in the best friendships. Something like Aristotle's virtue friendship reemerges as the ideal friendship. Though in some cases friendship between even the best people may present some moral danger, the best friendships are nonetheless those between virtuous people. Although friendship poses some moral danger, standards for excellent friendship depend upon the moral character of the friends.

In the first three chapters thus far, I have argued that friendship is best conceived of as a compound social entity composed of interdependent friends that, like an organism, requires the well-being of the parts that compose it in order to flourish. Doing so yields a variety of explanatory advantages when it comes to making sense of friendship, and it avoids committing us to making friendship a good in competition with moral goods more broadly. This does not mean that goods will never compete in particular cases; if one thinks that moral goods can end up in competition with each other, then it is no surprise that friendship also can end up in competition with other considerations. But this is not a unique problem for friendship as something that stands in tension with ethics, but rather a fact about the complexity of what we value much more generally.

With this account of friendship in hand, we are now in a position to see how considerations of friendship can arise in the development and use of social technologies. In Part II, I focus on situations where friendship is at issue in various applications of social robotic technologies.

Bibliography

Aristotle. 1999. *Nicomachean Ethics*. 2nd ed. Translated by Terence Irwin. Indianapolis, IN: Hackett Publishing Company. https://books.google.com/books?id=-Mf5XV8q6CgC&dq.

Brink, David O. 1999. "Eudaimonism, Love and Friendship, and Political Community." *Social Philosophy and Policy* 16 (1): 252–89. doi:10.1017/S0265052500002326.

Cocking, Dean, and Jeanette Kennett. 2000. "Friendship and Moral Danger." *The Journal of Philosophy* 97 (5): 278–96. doi:10.2307/2678396.

Cooper, John M. 1977. "Aristotle on the Forms of Friendship." *The Review of Metaphysics*, 619–48. doi:www.jstor.org/stable/20126987.

Foot, Philippa. 2002. *Virtues and Vices and Other Essays in Moral Philosophy*. Oxford Scholarship Online, Clarendon Press. https://books.google.com/books?id=fIHnCwAAQBAJ.

Jeske, Diane. 1997. "Friendship, Virtue, and Impartiality." *Philosophy and Phenomenological Research* 57 (1): 51–72. doi:10.2307/2953777.

Nehamas, Alexander. 2010. "XII—The Good of Friendship." *Proceedings of the Aristotelian Society* (Hardback) 110 (3pt3): 267–94. doi:10.1111/j.1467–9264.2010.00287.x.

O'Connor, David K. 1990. "Two Ideals of Friendship." *History of Philosophy Quarterly* 7 (2): 109–22.

Stocker, Michael. 1981. "Values and Purposes: The Limits of Teleology and the Ends of Friendship." *The Journal of Philosophy* 78 (12): 747–65. doi:10.2307/2026245.

Thomas, Laurence. 1987. "Friendship." *Synthese* 72 (2): 217–36. doi:10.1007/BF00413639.

Vlastos, Gregory. 1973. "The Individual as Object of Love in Plato." In *Platonic Studies*, 3–42. Limited Paperback Editions. Princeton: Princeton University Press. https://books.google.com/books?id=Ej553BF2bNIC.

White, Richard. 1999. "Friendship and Commitment." *The Journal of Value Inquiry* 33 (1): 79–88. doi:10.1023/A:1004332416151.

Part II
Robots

4 False Friends and False Coinage

A Tool for Navigating the Ethics of Sociable Robots

1. Introduction

Social robots are those technologies designed or used to engage human instinctive social responses. Part of the promise of these robots is that by engineering them to work with already-present norms and expectations for interaction, naïve users will find them more intuitive to use than those that require technical training or special vocabularies. They can also use these social appearances to address specifically social needs of users, from a desire for companionship to training in conversational skills. But part of the peril is that these social appearances, by stimulating deep-seated emotional and psychological reactions, can expose users to a variety of risks and harms.

Part I was devoted to articulating a theory of friendship. In Part II, we turn to our first social technology where a theory of friendship will be relevant. Engaging with ethical issues involving social robotics will both draw on theoretical resources already developed, and offer opportunities for refinement. Investigating the details of real-life situations involving novel technologies can help us to better understand what we value, by offering new opportunities to explore the boundaries of concepts, investigate the psychological effects of new sorts of interaction, and integrate social concerns with design and use considerations.

My goal in this first chapter on robots is to introduce an analogy from Aristotle that can help navigate issues involving the value of appearance in friendship, as they arise in social robotics. In the following chapters of Part II, I then show how the analogy can be used to make sense of several complex considerations likely to arise in the real-life implementation of these robots.

The roboticist Cynthia Breazeal defines a sociable robot as "socially intelligent in a human-like way, and interacting with it is like interacting with another person. At the pinnacle of achievement, they could befriend us, as we could them" (Breazeal 2003, 1). This might sound like science fiction, when read with a relatively strong conception of what it means to be "socially intelligent in a human-like way" and what it takes to be a friend. But herein lies the controversy. Social roboticists such as Breazeal focus on creating robotic systems that respond to human expressions, eye gaze, and tone, in ways that make them intuitive to the naive user, because their inputs tend to be the sorts of things to which we expect other human beings to attend. But this is a thin rather than a thick conception of human-like social intelligence. These are not the sorts of creatures we expect to enjoy rich inner lives, to value relationships in the ways described in the previous section, or require roboticists to develop artificial consciousness. Instead, her work and the work of others in social robotics focuses, specifically, on creating robots that work with our tendency to anthropomorphize, acting and reacting in ways that are consonant with our emotional systems. John Sullins calls this "affective design", which he says "seeks to imbed the robot deeply into the lifeworld of the humans with which it interacts. These machines are built to elicit, and even 'experience' emotion, in order to bond more fully with their human users" (Sullins 2008, 146). That is, this is not a far-future scenario in which Data from *Star Trek: The Next Generation* (Roddenberry 1987) works and socializes side-by-side with us, but rather an ongoing project to build robots within the limits of current technology that work with *our* pre-existing emotional and social processes. As a result, if we are to take Breazeal seriously, we have to read her as intending a relatively thin conception of friendship, as well as social intelligence. To those, like me, who think that friendship (as typically valued by human beings) is a more thick and rich phenomenon, it would be more perspicuous to say that the sorts of robots she envisions could behave *as if* befriending us, and that we could likewise react *as if* we were befriending them. Is this a distinction without a difference, or does it have ethical import? This last question is rapidly moving from the academic to the realm of the immediate and practical.

As robots move from the factory floor to the home, work, and healthcare sectors, non-technical users are increasingly likely to interact with them. Social robotics, broadly conceived, are designed to facilitate these interactions by designing and programming them so as to work with our existing social instincts, conventions, and emotional cues. From the cheerful song that the floor cleaning robot Roomba plays at the end of its cleaning cycle, to the humanoid robot NAO that is being utilized in therapies for children with autism spectrum disorders, to the robot seal Paro that is used as a companion for geriatric residents of long-term care facilities, we can find a wide variety of examples of such social integration strategies. And, notably, these do not posit the existence of robot consciousness, or robust robot agency, or the need for anything like robot rights. These are relatively simple, deterministic machines, more like cell phones than domestic dogs or cats when considering the level of cognitive complexity involved in their circuitry. What is novel and interesting about them is the way designers are working to capitalize on our anthropomorphic tendencies in order to encourage us to integrate them into our social fabric.

Of course, our ability to anthropomorphize even ordinary non-social robots and other utilitarian objects is impressive. A recent viral video shows a young child who encounters a water heater with a panel featuring two dials on top that look (to her) like eyes. The child responds to it by greeting it: "Hi robot! I love you, robot!" and hugs it affectionately (*Rayna Meets a "Robot"* 2017). Nor are emotional attachments limited to the young or ill-informed: soldiers on the battlefield have offered military funerals for destroyed bomb-detecting robots, and Matthias Scheutz relays the story of a field sergeant who called off a demonstration of a landmine-clearing robot that destroyed its limbs in the process, declaring it "inhumane" (Scheutz 2012). This suggests that the issue of robots' ability to engage our social responses and emotional systems cannot be avoided simply by discouraging design and implementation of explicitly and intentionally social robots. It also means that we ought not to wait until robust AI and artificial consciousness are on the scene in order to think about the ethics of robot/human interaction. Instead, we need ethical tools and frameworks that will deal with the issues we are already encountering.

Two such frameworks are offered by Mark Coeckelbergh and John Sullins. Coeckelbergh and Sullins both approach ethical questions by investigating how robots interact with humans, in particular by evaluating the design elements that affect human experience. This, they argue, is the most ethically relevant aspect of the current trend in domestic/consumer robotics. This takes us away from an older approach to robot ethics. Previously, many robot ethicists focused on what we might think of as *internal* features of robots. Coeckelbergh characterizes this traditional approach as follows: "the emphasis is . . . on the robot and what the robot really is or 'thinks', instead on how robots appear to us, humans" (Coeckelbergh 2009, 218). Such internal features had to do with conditions under which robots could be said to be conscious, to be moral agents, to be deserving of rights, and so on—essentially, questions about when and whether robots could be people. Coeckelbergh says that "this focus on the morality of robots means that ethical questions concerning how humans interact with robots and how humans experience that interaction remain out of sight" (2009, 218). Such a narrow focus is problematic both because it overlooks current areas of potential concern in human-robot interactions (because current robots fail to instantiate the relevant internal states) and because it "leaves out broader ethical questions, such as which lives we want to live (with and without robots)" (Coeckelbergh 2009, 218). This does not mean that humans are ignored altogether, but by focusing primarily on the robotic aspect of human/robot interactions, "[i]t limits ethics to concerns about things that might go wrong in interactions with robots; it leaves out the question: what if all goes right, is it still good to live with these robots?" (2009, 218). His approach rectifies these deficits by shifting from what he calls an internalist ethics to an externalist one (although he does not mean by this internalism and externalism about reasons, a common distinction in moral theory). Rather than focus on the internal states of robots and how these fit into extant moral frameworks about agency, responsibility, and moral status, we should focus on the external appearances of robots and how these engage our social and emotional responses, using theories that offer guidance on the sorts of social and emotional states it is good for *us* to enjoy.

Sullins (2008) also emphasizes the ethical importance of affective design, and the need for a framework that helps us to more clearly assess what robot appearances do to us, without getting sidetracked by questions about when, if ever, robots will achieve robust personhood. He also contributes an additional reason to take seriously the ethics of appearance, as distinct from questions about the robot mental life. He notes that some of the focus on the internal states of robots seems to be the result of Western cultural biases. Throughout much of the West, mind-body dualism of the roughly Cartesian variety enjoys widespread popularity. Meanwhile, many East Asian roboticists focus on issues of external appearance, and domestic robots' widespread popularity throughout East Asia does not seem to be encumbered by concerns about whether such robots are "really" conscious moral agents. Sullins notes that some of this difference of focus may be due to what Westerners often characterize as animist attitudes prevalent in Japanese and other East Asian cultures. Other features of worldview may also be relevant. For example, in *The Buddha and the Robot*, the Japanese roboticist Masahiro Mori (Mori 1981) makes the case that robots, like all things, partake in Buddha-nature simply in virtue of participating in causal interactions with the rest of the world. Buddhist ethical theories and theories of mind oppose precisely the sort of robust conception of a non-material self associated with Descartes, and in fact makes conscious-self questions non-issues or even mistakes for humans to worry about, let alone robots. That is, Western cultural biases may be distorting our perception of ethical issues involving robots. Especially given cross-cultural markets for robots, it might be wise to broaden our scope of inquiry.

Both Coeckelbergh and Sullins argue that ethicists should focus on robots' impact on human goods, rather than restrict ourselves to robot rules for action, or the quality of robots' internal experience. Given that we think some emotional responses and relationships are good and valuable parts of humans' experience, we should be interested in how robots can constructively contribute to these. Coeckelbergh explicitly calls this a shift from an ethics of reality to an ethics of experiences.

> I suggest that we start from studies of how humans interact with robots on the basis of apparent rather than real humanoid features

> (intelligence, consciousness, emotion, etc.). For example, does the face of a particular robot appear human, and if so, how do we experience this feature in interaction with that robot? What robots do to us, depends on how they appear to us, not on what is 'really' in their mind. For instance, existing robots are not sentient and lack feeling; nevertheless, when humans interact with some types of robots they may act and talk as if the robot has sensations and feelings. In a similar way, humans tend to attribute thoughts and beliefs to robots. These observations are relevant to the way we humans act and live. Rather than focusing our ethical worries on robots, let us worry about humans, about what we think, feel, and dream of.
>
> (Coeckelbergh 2009, 219)

But this can take us too far in the other direction. What if what humans care about *is* what, if anything, is going on in the robot's mind? Coeckelbergh says, "I remain agnostic with regard to any theory of robot 'mind', since I hold that—at least when applied to the robot—such theories of mind are not very relevant to the ethical aspects of human-robot interaction" (2009, 220). But once we start thinking about relationships, surely what other parties to a relationship think (or don't think) is highly relevant to many people.

Now, one might think he is asking about something other than robot/human relationships, such as the effect interactions with robots may have on human character. But he explicitly characterizes his framework in terms of relationships and their role in human good:

> Can human good appear in human-robot interaction (or relationships), or only in human-human interaction (and relationships)? Can human-robot interaction (relationships) contribute to human flourishing and happiness? Can such interactions constitute friendship, love, or relationships at all? Can they co-shape a flourishing community?
>
> (Coeckelbergh 2009, 220)

To answer these questions, I suspect we cannot dismiss the issue of robot mental states so easily. This does not mean that we should wait for robot consciousness to engage with the issue of robot friendship. But granting that today's robots do not have anything very much like a human mental life, this fact itself seems relevant to ethical questions about social robots.

Perhaps in deference to such questions, Sullins proposes a more moderate approach to an ethics of robot appearances, one that does not dismiss the relevance of their (lack of) internal states altogether. He asks, "How will we program these machines to interact with us as friends?" (Sullins 2008, 144). Part of his answer involves affective design, which "seeks to imbed the robot deeply into the lifeworld of the humans with which it interacts. These machines are built to elicit, and even 'experience' emotion, in order to bond more fully with their human users" (2008, 146). This does not fundamentally change their nature, but does change how we experience them. "[S]uccessful affective robots", he says, "will be machines that are designed to do what machines do best, but in a way that engages the users' natural anthropomorphizing tendencies to help embed that machine in the user's lifeworld". Given our human focus on sociality, this "means that affective robots are best when they elicit our natural human predispositions to grant personalities to the objects around us making it easier for us to interact with the technology" (2008, 148). But unlike Coeckelbergh, he voices specific concerns about the consequences of activating our tendency to "personalize" robots. In his conclusion he calls for the development of robots that are anthropomorphic and utilize our emotional responses to more effectively and artfully interact with technology, but cautions that they ought to remain stylized or cartoonish to avoid deception. Furthermore, he says they ought not to replace human interactions. He does not say much, in this piece, to explain why these concerns are relevant, but we might imagine that this is because humans have historically valued some things about human/human interactions that robot/human interactions cannot be expected to duplicate, and furthermore that these things cannot be experiential in character.

This approach is consistent with the first of two ways that Coeckelbergh identifies to evaluate robot ethics given a move from robot-centered frameworks to human-centered ones: "One approach is to start from a certain conception of human good, of human flourishing, of happiness, of friendship, and of love" (2009, 220). But Coeckelbergh worries that this will be inadequate to the task, as pre-existing conceptions are formed in situations that are by hypothesis devoid of precisely the sorts of creatures we now face, and so cannot be informed by evolving practices or values that include them. This leads him to the approach he prefers: "Rather than starting from the capabilities list as *a priori* moral norms, we need to start from concrete experiences and imagination of human-robot interaction and then discuss what good understood in terms of capabilities means" (2009, 220). The worry here appears to be that some people simply may not care about the internal lives of robot friends in the way that they care about the internal lives of their human friends. To pre-judge these people as mistaken would be to impose a code of norms on them that might not be appropriate to their way of life. "Good", he says, "is not independent of what happens in practice; it can only exist and flourish in practice" (2009, 220). So, rather than approach ethical questions using preconceived notions of sociality and emotional well-being, he argues, "[l]et us listen to people's experience and use our moral imagination to find out if there are possibilities of living with robots that enhance human flourishing and happiness" (2009, 220). That is, "we must be attentive to, and imagine, possibilities of living with personal robots that contribute to, and indeed co-constitute, good human lives in practice" (2009, 221). The fact that people "personalize" robots is not necessarily something we need to guard against via "cartoonish" or stylized design, and the difference between robots and humans need not remain explicit.

This appears to be the difference between a top-down and bottom-up approach to an ethics of human-robot interactions and affective design. Sullins is engaging in something like a top-down approach, starting from established conceptions of the nature and value of human sociality, and reasoning from there to defend affective design only insofar as it exercises emotional abilities that ultimately facilitate the (already-of-settled-value) human relationships. Coeckelbergh is advocating a wait-and-see approach, wherein we determine ethical prescriptions only after observing human values in practice in relevant contexts. The potential drawback to Sullins's approach is that it may miss important emerging or context-specific factors not already built into pre-existing conceptions of human good. The potential drawback to Coeckelbergh's approach is that human practices do not always mesh terribly well with our most central values—sometimes we say and do things, even habitually, or even profess to value things, which do not reflect our all-things-considered values. We can get things wrong in how we live our day-to-day lives, and one of the valuable things about working from something like an established conception of the good is that it helps give us a standard against which to check our practices and activities, to help us to live with integrity. Even when we are willing to adjust our guiding concepts and principles when they prove fundamentally incompatible with lived experience, they can be useful tools in figuring out how to live our lives, and the mere fact that they *sometimes* clash with some experiences seems insufficient reason to abandon using them altogether.

I am sympathetic to Coeckelbergh's concerns, but I think it would be going too far to focus *exclusively* on appearances, especially when the concept at hand (friendship) already seems shot through with concern for appearances versus reality (as when we wonder who our "true" friends are). It would be too quick to unthinkingly declare that friends *must* be human, simply because they always have been in the past. But if introspecting on what we value about friendship leads us to conclude that robots cannot give us all that we value about it, this seems an important finding for ethical prescriptions about the use of robots for *us*. In what follows, I lay out an account of the role of appearances versus reality in ethical issues involving friendship. In the following two chapters, I use this tool to evaluate two paternalistic uses of social robotics, in which we must decide on behalf of cognitively challenged patients what will most help them to live good lives, when they are not in a position to do so. In the last chapter in this part, I return to the more difficult question of what we should think when others' conception of friendship diverges in the way Coeckelbergh suggests, where appearances seem "good enough" to some (but not all).

2. Ethical Problems of Sociable Robots

The use of robots to satisfy social needs strikes many as problematic. The challenge is to explain why this could be so without presupposing that we have already figured out all there is to know about human sociality.

In what follows, as elsewhere in this volume, I draw on the Aristotelian tradition to make sense of this intuition. Aristotle's eudaemonist theory of virtue strikes many as a promising way to approach ethical questions involving emerging technologies. It invites us to consider what constitutes a good, flourishing life, and the impact of both large and small actions and influences on our ability to live such a life (Bynum 2006; Vallor 2011).

At the start of his discussion of friendship in the *Nicomachean Ethics*, Aristotle argues that friendship is valuable intrinsically, in its own right, not merely as a means to other ends: "no one would choose to live without friends even if he had all the other goods" (Aristotle 1999, 1155a5). Thus, the fact that sociable robots present significant instrumental benefits (as they surely do) could not, even in principle, justify them as meeting our social *needs* in the richer sense of providing us with the ingredients for a good life. That is, we should not expect the instrumental value of sociable robots to completely justify them in their capacity as companions. If they are to be justified, it will have to be by other means.

A thought experiment can be used to show that, even were one to develop a robot with which interactions were indistinguishable, to the patient, from those of a human companion, such a creation would not suffice to provide the full goods of friendship.

Imagine that you are given a choice between two possible lives. You know at the time of the choice how the lives will differ. But once you make your choice and begin your chosen life, you will forget ever having been given the choice—it will be as though this is the way things have always been. In one option, the people you consider your closest friends are actually paid actors, although if you were to choose this life, once it commenced you never learn of their illusory nature, and could never observe it from their actions. These friend-facsimiles would not use the appearance of friendship to exploit you. They would not betray your confidence, or use your trust to take advantage of you, but they would also not care for you or find pleasure in interacting with you. Call this the Truman Show option. In the other world, your closest friends are exactly as they appear to you to be. Call this the Genuine option.

Given the choice, most of us would prefer the Genuine over the Truman Show life. This shows at least two things. First, that we value more than appearances in friendship, because one is preferable to the other and yet both provide the appearance of friendship. Second, that the Genuine option is choice worthy for non-instrumental reasons. In Truman Show, one gets all the instrumental benefits of friendship, because the friends outwardly act and provide all the same external goods as in Genuine. Furthermore, they cause none of the harms ordinarily associated with "false friends". And yet it is less attractive than Genuine. So whatever is good about Genuine cannot be the instrumental benefits and costs. Our preferences are not based on differences in experiential quality, nor in the bad outcome of having actors as "friends". This thought experiment shows that the most choice worthy lives involve genuine friendships, which involve reciprocal caring of genuine agents capable of care. Such relationships must be intrinsically valuable.

So when we assess the value of sociable robots, it is not sufficient to show that patients accept them as companions to show that they provide us with the goods of sociality. (As we will see in the next chapter, dementia patients presented with an AIBO thought it was a baby, but their enjoyment found in such false beliefs is not necessarily sufficient to justify deception.) If the patients are unaware of the nature of the things from which they derive social satisfaction, they may not be in possession of the good they take themselves to have, even when they can be observed to reap the instrumental benefits of sociality. Unless the robots themselves are the sorts of things we can conceive of as agents who value us and take pleasure in our company, as the genuine friends do in Genuine, we run the risk of selling patients a false bill of goods, leaving them worse off than they think they are. This seems a high ethical cost of using sociable robots, even though it is not applicable in every case.

I agree with Vallor (2011) that in such matters, we should not look for a uniform principle to guide us. Details of context matter, and identifying relevant features of context is a nontrivial task (Misselhorn, Pompe, and Stapleton 2013). Complicating matters further, the varying cognitive, emotional, and volitional abilities of patients for whom such technologies are prescribed may influence what counts as a good choice for a given subject. Some users recognize that they are interacting with robots, while others, owing to various kinds of cognitive disorders or features of context, do not.

In evaluating the well-being of users, a broad and rich conception of well-being should be utilized. It should not be limited to thin, easily-operationalized concepts like subjective pleasure or physical health markers, but should also include whether these patients are living good lives of the sort we consider choice worthy; whether their desires are really being met or only appear to be satisfied. Instead, something more eudaimonist, which includes both physical and psychological health, as well as richer and less easily quantified goods, would be desirable.

In addition to the well-being of users of social robots, the impact of sociable robots on caregivers, designers, and users' families and friends should be taken into consideration (Vallor 2011). Using robots in elder-care, for example, may help with inequalities reinforced by care work (Parks 2010). But there are also costs. If using robots to satisfy elderly people's social needs causes us to devalue the elderly, this seems to count as a reason against their use even if the patients themselves do not feel the impact of this effect. If substituting robot interactions for human ones tends to weaken family ties, that could also affect people beyond the individual user. If using robots to deceive patients into thinking they have friends when they do not makes us more comfortable with deception, we have reason to tread with caution in their design and implementation.

3. Of Economies and Friendships

An analogy from Aristotle's discussion of deception and friendship provides a resource of the kind described above. But before introducing it, we should get clear on what exactly friendship consists in. The thought experiment involving Truman Show (Weir 1998) and Genuine options appealed to some intuitions on this front, but a more careful and clear articulation will prove useful in understanding what deception about friendship consists in.

As we saw in the previous section, we have reason to take friendships to be objects in our social ontology—a kind of very small and close-knit group. Doing so offers a way to make sense of some common intuitions about friendship both within and without the Aristotelian tradition. It can accommodate both shared identity/unity and the importance of complementary differences.

Now, suppose one thinks of friendships and other close-knit social groups as objects in one's social ontology. How does this apply to social robotics? If we construe friends as parts of composite objects—friendships—we can explain their shared identity in terms of parts of a whole rather than similarity between the friends. Inter-responsiveness and interdependence, as features of parts jointly composing a whole, then come to the forefront as characteristics of friends. This is consonant with many ordinary beliefs about characteristic qualities of friendship: that true friends reciprocate, that friends are characteristically those whose emotional states are responsive to their friends' well-being, and that friends' interests, broadly construed, include both their own well-being, that of their friends, and the wish that the two should remain interdependently connected.

Note that this does not preclude the possibility of similarities between friends, nor does it rule out similarity as one possible way that inter-responsiveness and interdependence can emerge and be sustained in friendship. It does, however, shift our focus, in thinking about friendship, from narrow consideration of the features of individuals, to features of the social groups they compose.

It has the explanatory advantage of accounting for the difference in intuitions between the Truman Show case and reciprocal caring, as the difference between real and merely-apparent social phenomena. In Truman Show, one is not appropriately interdependent with one's friends: the dependency runs only one way, as care and emotional responsiveness run in only one direction. In Genuine, by contrast, friends are mutually responsive to each other as parts composing a whole—there exists a genuine friendship, and not merely the appearance of one. Recall that according to Aristotle, friendships have intrinsic value. This interpretation supports the intuition that friends are not valuable merely for the experiences or other instrumental goods they provide.

I now apply this theory of friendship to unpacking an analogy Aristotle offers, in order to better understand concerns about the potential badness of some uses of sociable robots to provide the subjective appearance of friendship without the existence of a grounding entity. In Book IX of the *Nicomachean Ethics*, following a discussion of friends as other selves and a detailed exploration of the reasons that conflict arises in various kinds of friendship, Aristotle comments on problems that arise between people when appearances of friendship fail, in various ways, to match reality. About this, he says the following:

> We might . . . accuse a friend if he really liked us for utility or pleasure, and pretended to like us for our character . . . if we mistakenly suppose we are loved for our character when our friend is doing nothing to suggest this, we must hold ourselves responsible. But if we are deceived by his pretense, we are justified in accusing him—even more justified than in accusing debasers of the currency, to the extent that his evildoing debases something more precious.
>
> (Aristotle 1999, 1165b)

It is worth noting that this account explains the badness of false friends not in terms of malicious intent, but representation of one thing as another. This is consistent with intuitions about the Truman Show scenario, where paid actors seem poor substitutes for friends even if they never betray one's trust or seek to use this illusion to harm the person they "befriend".

Rather, Aristotle's analogy between false friends and false coinage suggests that money constitutes membership in an economy. An economy, like a friendship, is a social group, albeit of a different kind. One can speak, for example, of the way the British economy reacts to a war, because an economy, like a friendship, is dependent upon and partly defined by the interdependence and inter-responsiveness of its members. Like friendships, economies derive their value from this. Unlike friendship, the value of an economy appears to be primarily instrumental, and one might opt for a world without economies if the same external goods were equally well realized by other means, unlike in friendship. But the analogy need not be perfect in order to be instructive. False money gives a false impression of membership in a social group, and its badness derives from this specific misrepresentation; likewise for friendship.

This suggests that there is something independently bad about counterfeiting, over and above the harm any given individual may or may not suffer in handling counterfeit currency. Consider, by way of illustration, the Case of the Compassionate Counterfeiter:

> *Compassionate Counterfeiter:* A is experiencing anxiety about money, which would be alleviated if A believed A possessed more of it. Out of concern for A, in order to assuage these worries, B writes A a bad check. As it happens, A never deposits the check, and eventually receives (actual) funds sufficient for financial support from a new job.

It seems to me that B has done something wrong in writing the bad check, even though A never cashes it. B's wrongdoing seems to consist in giving A the impression that A has economic connections that A could draw on, but which A does not in fact have. Economies, unlike friendships, are primarily instrumentally valuable, which may change the picture somewhat. Different goods may be implicated in the false appearance of each. But insofar as both friends and money constitute membership in a social group which people find valuable, the false appearance of membership seems bad in similar ways.

This is not meant to imply either that people are not harmed, or have no reason to complain, when they are given the appearance of friendship without the reality. However, it suggests that the harm they suffer may be explained by the specific kind of deception: being given the impression that they are involved in something of value when there is nothing in which to be involved. We know, from the discussion so far, that we value friendships intrinsically. We have reason to believe that certain kinds of cognitive confusion, prevalent in those very populations toward which some social robot designs are directed, in conjunction with our common vulnerability to certain features and gestures that elicit strong and often involuntary emotional reactions (such as big "eyes" and other anthropomorphic features), can cause the false appearance of friendship. We ought, then, to avoid exploiting these vulnerabilities in order to produce counterfeit friends that fool people into thinking they have something with a particular kind of value that is really absent.

4. Comparing Aristotle's Analogy to the Competition

Aristotle's analogy directs us to consider both whether people are fooled into thinking they are members of social groups, and who is to blame for the deception. This is helpful in identifying potentially relevant features of context in assessing the ethics of a given situation. The analogy introduces an account of the badness of sociable robots—not always an overriding badness, but a reason to be judicious in their design and use. It is helpful to consider how this account compares with other extant accounts of said badness, which can be thought of as attempts to explain the intuition that sociable robots can be ethically problematic. In what follows, I compare these accounts of badness in order to show why the version suggested by Aristotle's analogy is superior to others on offer.

Some accounts focus on the putative badness of sociable robots' capacity for enchantment. They hold that because sociable robots "enchant" by appealing to social instincts, they thereby deceive us, even against our better judgment (Grodzinsky, Miller, and Wolf 2015; Turkle 2011). However, the proponent of an Aristotelian counterfeiting analogy can respond that this seems problematic. First, it makes seemingly self-aware testimony of people who report benefits of interaction with sociable robots to be unreliable, which runs the risk of inappropriate paternalism. This seems to be the sort of prejudgment Coeckelbergh worries about—perhaps the fact that people enjoy the experience of interacting with robots *as if* they possessed genuine emotions, while knowing that they are not, just shows the inadequacy of some extant accounts of the value of social interaction. Respect for agency, and for people's own judgments about what constitutes their good, ought to be part of responsible robot ethics. We ought not to override people's expressed preferences simply because they disagree with ours. Secondly and more generally, it is implausible to think that "enchantment" is sufficient for badness: we voluntarily watch "tearjerker" movies and read comedic novels, both of which work by appealing to our social responses and eliciting powerful emotional reactions, despite our intellectual judgments that the characters in the story are not real. Such "deception" seems wholly innocent. More broadly, playful simulacra of real phenomena can enrich our lives. Stories engage our social responses. Monopoly "money" can engage our economic reasoning faculties. This does not mean that we can never err by retreating into fantasy as a more enjoyable alternative to reality. But even when enchantment is bad, it may sometimes be plausible that blame should fall on the user, as Aristotle cautions us, and does not necessarily show that anything is wrong with the object. The fact that some patients may misjudge their own social needs is not, in itself, reason to prohibit use of sociable robots altogether. Consider that people sometimes make ill-advised choices when it comes to friends and companions; this does not license others to run their social lives. Enchantment thus seems an unsatisfactory explanation of the badness of sociable robots.

Another kind of account focuses on robots' potential to substitute for human relationships as the ultimate source of their badness. "Substitution" accounts say sociable robots are bad when and because they substitute for interaction with human beings (Sharkey and Sharkey 2010; Sorell and Draper 2014). But the counterfeiter can respond that it is an empirical question whether robots will substitute for human interaction or facilitate it (Vallor 2011). For instance, Wada and Shibata found that introducing Paro to nursing home residents increased both the quality and quantity of patients' interactions with each other, including a marked reduction in what they characterized as "backbiting" (Wada and Shibata 2006). This is not so surprising if we consider that social abilities may, like muscles, get stronger with practice. It may also be that, by using robots to address some of the more stressful parts of caregiving, for example, people will find it more enjoyable (and hence be more likely to make it a priority) to spend time with friends and relatives who require assistance. It is not wildly implausible to think that people who are less depressed and more stimulated will be more pleasant and less draining to be around, thus creating a virtuous circle whereby they are included in more social events, and hence less likely to suffer from the ill effects of loneliness.

It is also worth noting that not every substitution of robot for human interaction is bad. After all, not all human interaction is desirable (Sorell and Draper 2014), and robots may supplant interactions with abusive or insensitive caregivers, or even just meddling busybodies, people whose presence in a person's life would not constitute a good (Vallor 2011). But the substitution account faces another disadvantage: even without causing a decrease in human interaction, it seems intuitively problematic to mislead patients about the nature of what they are interacting with, and the substitution account cannot explain why.

This last intuition motivates a "deceptiveness" account of the badness of sociable robots. According to a generic deceptiveness account, sociable robots may be bad when they deceive people, because they deceive people. Benign deceptions (defined as good intentions plus good consequences) are possible but rare exceptions to this principle (Grodzinsky et al., 2015). Because the counterfeiting account is a kind of deceptiveness account, it may seem odd for me to object to such accounts. But the generic deceptiveness account differs from the counterfeiting account in its level of generality. Deception is often bad, but the times when it is least problematic are precisely the sorts of cases for which many sociable robots are designed. For instance, deceptiveness may seem—in some cases—permissible for paternalistic reasons, as when cognitive confusion makes a person a poor judge of their own good. As Grodzinsky et al. note, it may also be justified when done to facilitate ease of use, as when a Graphical User Interface (GUI) designer calls something a "file folder" even though no physical folders are involved. One might think that in some cases, transparency would impede the robot fulfilling its intended function, especially with a population not familiar with cutting-edge technology, such as geriatric care robots in nursing homes. It would not be practical to give seniors a crash course in robotics before using robots for basic care. Metaphors can help non-technical users to successfully interact with advanced technologies.

Because we need to distinguish permissible from impermissible deception, and because there are instrumental benefits to deploying sociable robots, as detailed earlier, more detail is required. Clarifying the content of the deception can help us better evaluate individual tradeoffs, and the counterfeit account provides such clarification. Given that sociable robots can provide patients with the subjective experience of friendship, we cannot yet determine whether or not this constitutes a benign deception. Rather, we need to ask, is giving someone the subjective experience of friendship a good thing to do? This is what the counterfeit account addresses, by clarifying the importance of making sure the experiencer understands the nature of the source of the experience.

5. Conclusion

Aristotle's analogy between counterfeit currency and false friendship thus provides guidance on the ethical use of sociable robots, without claiming to offer a uniform solution that glosses over important details of context. By characterizing the badness of sociable robots (when there is badness to be characterized) as giving the false impression of membership in a non-existent social group, we can minimize the risks of said harm. There remains the possibility that the harms of counterfeiting friendship may be outweighed by other considerations, just as other kinds of deception might be justified in some circumstances. Some benefits associated with sociable robot use are quite powerful, as we will see in the next two chapters. But this seems an advantage rather than a drawback of the theory. Living good lives and helping others to do so, especially those whose cognitive and physical disabilities might significantly impair their ability to live good lives (which, as we will see in the next two chapters, applies to many current users of social robotics), is a complex and demanding problem that will occasionally require tradeoffs between competing goods. Nonetheless, the analogy remains useful in helping us to correctly evaluate what goods are in competition, and when.

In the following chapter, this analogy will be put to use evaluating the ethics of companionate robots for geriatric patients. Here, questions of cognitive impairment, confusion, and loneliness will be quite salient, while at the same time emerging empirical data supports the potential health value of exercising social responses via robotic means.

Bibliography

Aristotle. 1999. *Nicomachean Ethics*. 2nd ed. Translated by Terence Irwin. Indianapolis, IN: Hackett Publishing Company. https://books.google.com/books?id=-Mf5XV8q6CgC&dq.

Breazeal, Cynthia. 2003. "Emotion and Sociable Humanoid Robots." *International Journal of Human-Computer Studies* 59 (1–2): 119–55. doi:10.1016/S1071–5819(03)00018–1.

Bynum, Terrell Ward. 2006. "Flourishing Ethics." *Ethics and Information Technology* 8 (4): 157–73. doi:10.1007/s10676-006-9107-1.

Coeckelbergh, Mark. 2009. "Personal Robots, Appearance, and Human Good: A Methodological Reflection on Roboethics." *International Journal of Social Robotics* 1 (3): 217–21. doi:10.1007/s12369-009-0026-2.

Grodzinsky, Frances S., Keith W. Miller, and Marty J. Wolf. 2015. "Developing Automated Deceptions and the Impact on Trust." *Philosophy & Technology* 28 (1): 91–105. doi:10.1007/s13347-014-0158-7.

Misselhorn, Catrin, Ulrike Pompe, and Mog Stapleton. 2013. "Ethical Considerations Regarding the Use of Social Robots in the Fourth Age." *Geropsych* 26 (2): 121–33. doi: 10.1024/1662-9647/a000088

Mori, Masahiro. 1981. *The Buddha in the Robot*. 1st English ed. Edited by Ralph Friedrich. Translated by Charles S. Terry. Tokyo, Japan: Kōsei Publishing.

Parks, Jennifer A. 2010. "Lifting the Burden of Women's Care Work: Should Robots Replace the 'Human Touch'?" *Hypatia* 25 (1): 100–20. doi:10.1111/j.1527–2001.2009.01086.x.

Rayna Meets a "Robot". 2017. YouTube. Columbia, MD. https://youtu.be/h1E-FlguwGw.

Roddenberry, Gene. 1987. "Star Trek: The Next Generation." NTSC. *Episodes 1–176*. Paramount Television.

Scheutz, Matthias. 2012. "The Inherent Dangers of Unidirectional Emotional Bonds between Humans and Social Robots." In *Robot Ethics: The Ethical and Social Implications of Robotics*, edited by Patrick Lin, Keith Abney, and George A. Bekey, 205–22. Cambridge, MA: MIT Press.

Sharkey, Amanda, and Noel Sharkey. 2010. "Living with Robots: Ethical Tradeoffs in Eldercare." In *Close Engagements with Artificial Companions: Key Social, Psychological, Ethical and Design Issues*, edited by Y. Wilks, 245–56. Natural Language Processing. Amsterdam: John Benjamins Publishing Company. https://books.google.com/books?id=EPznZHeG89cC.

Sorell, Tom, and Heather Draper. 2014. "Robot Carers, Ethics, and Older People." *Ethics and Information Technology* 16 (3): 183–95. doi:10.1007/s10676-014-9344-7.

Sullins, John P. 2008. "Friends by Design: A Design Philosophy for Personal Robotics Technology." In *Philosophy and Design*, 143–57. The Netherlands: Springer. http://link.springer.com/chapter/10.1007/978-1-4020-6591-0_11.

Turkle, Sherry. 2011. *Alone Together: Why We Expect More from Technology and Less from Each Other*. New York, NY: Basic Books. https://books.google.com/books?id=J2ine5sIIkgC.

Vallor, Shannon. 2011. "Carebots and Caregivers: Sustaining the Ethical Ideal of Care in the Twenty-First Century." *Philosophy & Technology* 24 (3): 251. doi:10.1007/s13347–011–0015-x.

Wada, Kazuyoshi, and Takanori Shibata. 2006. "Robot Therapy in a Care House—Its Sociopsychological and Physiological Effects on the Residents." In *Proceedings 2006 IEEE International Conference on Robotics and Automation, 2006. ICRA 2006.*, 3966–71. doi:10.1109/ROBOT.2006.1642310.
Weir, Peter. 1998. *The Truman Show.* 35 mm. Paramount Pictures.

5 What's Wrong With Robot "Friends" for Lonely Seniors?

1. Introduction

I am generally optimistic about emerging technologies, but at least one use of social robotics gives me pause. This chapter expresses my worry about it while recognizing some tangible benefits to its responsible use. In the last chapter, Aristotle's analogy between false friends and false coinage was introduced as a general tool for thinking about social robotics. Here, the analogy is used to make sense of ethical issues in social robotics for geriatric care. In this context, caregivers must make decisions that are, to the degree possible, consistent with patients' values and respectful of their beliefs while also advancing their well-being, a task that can be complicated by a variety of cognitive impairments that can occur and progress at varying rates among seniors. These impairments can make accidental deception a real concern and one that may vary day-to-day for individual patients.

Social robotics shows great promise in the field of geriatric care, where robots can be used to address a number of standing concerns. They can help offset staffing shortages (Sparrow and Sparrow 2006), which are of particular concern in places where the elderly constitute an increasingly large percentage of the population (a consequence of falling birthrates), as is the case in many parts of Asia, Europe, and the Americas. Robots can increase the autonomy of geriatric patients, helping them to perform tasks they would otherwise be unable to manage (physically or mentally) on their own (Borenstein and Pearson 2010). Robots, at least when properly designed and built, can provide a consistent quality of care that would be likely to constitute a significant improvement over the notorious inconsistencies of human geriatric caregivers (Vallor 2011). Robots can perform at least as well as human caregivers at some tasks. In fact, robots may be preferable to humans in circumstances such as assisting with bathing, and using the toilet, where seniors who value privacy and dignity may prefer to use non-human assistance rather than exposing themselves to even the best human aides (Sharkey 2014). Robots can be customized to address the particular needs of different patients. And they can be a constant presence for both housebound and institutionalized seniors where it would be cost-prohibitive to provide round-the-clock human assistance.

They thus show promise on a number of fronts.

The application in which I am interested here is the use of robots to meet seniors' social needs. This may overlap with a number of the implementations sketched above, but has also proved a compelling enough area that some robots have been developed specifically for this purpose.

Companionate robots are used to alleviate loneliness in geriatric patients. Their ability to do so derives from their apparent responsiveness and emotional appeal. But this introduces the threat of a false appearance of friendship, especially to patients suffering from cognitive impairments. This raises a number of questions, chief among them ones about quality of life: is a subjectively socially satisfying life with robotic companions a good one? Is it the sort of thing responsible caretakers can or should provide? To what extent does cognitive confusion of a patient introduce ethical costs to this practice? My project here is to apply the analogy between false friends and false coinage, discussed in the previous chapter, to help in developing answers to such questions.

2. Sociable Robots: Pros and Cons

Loneliness is a major problem in geriatric care. And loneliness impacts health. For example, alleviating loneliness can help slow the deterioration of cognitive faculties. "An extensive social network seems to protect against dementia", report Fratiglioni et al. (2000, 1315). Even infrequent social contacts reduce occurrence of dementia "if such contacts were experienced as satisfying" (Fratiglioni et al 2000, 1318). Risk of Alzheimer's disease is sensitive to "perceived isolation", with risk "more than doubled in lonely persons compared with persons who were not lonely" (Wilson et al., 2007, 234). Alleviating loneliness via subjectively satisfying experiences, including interactions with sociable robots, can also be beneficial in other respects: For instance, Wada and Shibata (2006) observed the results of introducing robots to elderly residents of a care house, and found that "urinary tests showed that the reactions of the subjects' vital organs to stress were improved after the introduction of the robots" (2006, 3966).

Many sociable robots are designed to address the tangible costs of loneliness. Among the most well-known of these is Paro, a robotic baby seal who has even been approved as a medical device by the FDA (Ponte 2014). Sociable robots have anthropomorphic features, such as big eyes and expressive faces, contributing to their ability to elicit appropriate emotional reactions. They are highly interactive; some respond to speech, others to gestures. Paro differentiates between light touch, such as stroking, and hard contact such as striking. It responds by adapting its activity to avoid provoking reactions like striking, which may indicate that something it has done has been upsetting to patients, and increasing the frequency of behavior that results in petting and scratching. They can thus tailor their behavior to reflect the preferences of particular users.

It is an open question whether seniors' needs will best be addressed by dedicated, specialized sociable robots, or whether these features will eventually be implemented in robots designed to serve other caregiving purposes as well. What I will argue here, although drawn primarily from the current generation of sociable robots, has implications for geriatric care robots more generally. Fortunately, Paro and other sociable robots have been in use for long enough that we have some data on their impact on geriatric patients. This allows us to say with some confidence what some of the issues presented by current and future iterations of sociable robots may be.

This new technology offers some distinct advantages to alleviating loneliness in geriatric patients. Robots have been shown to reduce patient loneliness with comparable rates to those of animal-assisted therapy, without the hygiene concerns of using live animals around patients with weak immune systems (Banks, Willoughby, and Banks 2008) or exposing animals to patients suffering from mood swings or impulse control problems. They have been successfully used with both lucid patients, and patients suffering from cognitive deficits. For instance, a group of researchers report patients with severe dementia

> recognized that AIBO [their robot] was a robot. However, once we dressed AIBO, the patients perceived AIBO as either a dog or a baby. Nevertheless, the presentation of AIBO resulted in positive outcomes for the severe dementia patients, including increased communication between the patients and AIBO.
>
> (Tamura et al., 2004, 83)

This example is striking because it shows that one need not be cognizant of the nature of the robot to be benefited.

Sociable robots can help buffer caretakers from some of the costs of geriatric care. Patients may experience mood swings, depression, tantrums, and emotional outbursts, problems often compounded when caregivers have long-standing, complicated histories with their charges, as when adult children care for elderly parents. Such histories can exacerbate emotional responses of both cared-for and caregiver, in ways that benefit neither. Caring for the elderly can be psychologically taxing, but sociable robots are not vulnerable to these costs. They may also help address feminist concerns about the "care burden" of (predominantly women's) caretaking labor (Parks 2010).

In addition to benefiting cognitively compromised patients, as noted above, lucid ones can also find them comforting. For instance, a

74-year-old reported, "The first time, I didn't like playing with the robot because I was depressed. After I had played with the robot several times, I felt good". Another said, "I do not think about anything while playing with the pet-type robot. It heals my mind" (Kanamori, Suzuki, and Tanaka 2002, 214). That is, at least some patients do not need to be fooled in order to experience benefits of interacting with sociable robots.

But surely the loneliness these seniors feel (and which the robots alleviate) is only the tip of an iceberg. Many are actually experiencing social isolation, not just the appearance of it. As seniors age, their mobility limits their ability to visit friends, and an aging cohort means many long-time companions are likely to have passed away. Families, often important sources of friendship on an inclusive conception of friends, are busy as children grow up and move on with their own lives. And even when professional caregivers are present, they are, for many reasons, unlikely to make good sources of the kind of sustained, particular, highly individualized emotional interdependence characteristic of friendship. Friendship seems to require emotional involvement, mutual caring, and mutual responsiveness. Furthermore, friendship seems to be an important good—a component of a choice worthy life, and so one with which we ought to be concerned when we concern ourselves with patients' quality of life.

Robots, then, pose two kinds of risk. First, that caretakers will mistake patients' alleviation of loneliness for satisfaction of a still-missing component of patients' well-being. And second, that seniors may be vulnerable to this same error.

"But is that really a mistake with which we ought to concern ourselves?" A skeptic might ask. "After all, if Grandma's content and thinks she has a friend, what harm is there in that?"

What follows is my argument that it is a significant mistake. My method here, as throughout this book, may be thought of as broadly eudaimonist, as I invite us to consider what constitutes a good life, and the impact of both large and small influences on our ability to live such a life, in order to better assess questions about quality of life and its attendant implications for responsible caregiving.

3. Previous Accounts

Extant accounts attempt to explain the inadequacy of sociable robots at meeting our need for friendship, but these accounts are problematic, as I will explain. Reviewing their shortcomings will help in developing a more successful theory.

Some worry that the use of robots for geriatric care, particularly those parts that seem most emotionally and ethically important (such as addressing our need for social contact) ends up devaluing (human) care and reinforcing "broader social attitudes towards older persons" as unworthy of our attention and effort (Sparrow and Sparrow 2006, 143). But this seems to depend on whether providing robot care actually counts as devaluing patients or human care, or improving on it, which will require us to say more about the value and cost of companionate robots for seniors to begin with. Because they elicit powerful emotional responses from users, who respond to them as if they were genuine agents, they are potentially deceptive (Sullins 2008; Grodzinsky, Miller, and Wolf 2015). As we saw in the last chapter, this is not an idle concern, but clarifying the nature of the deception can help us to maximize benefits while avoiding deception where it is possible or worthwhile to do so. We might worry that the use of simple, uncomplicated, appealing but quite "shallow" sociable robots provide short-term cessation of loneliness but ultimately decrease one's capacity for genuine connection in actual messy, complex, demanding human relationships (Turkle 2011). I will address this concern in Chapter 7. If the use of sociable robots becomes widespread, we might worry that this will increase *actual* loneliness of patients (as opposed to perceived loneliness). That is, if it becomes common to use robots to care for the elderly, this may contribute to the isolation of seniors, insofar as it leads us to substitute robot relationships for human ones (Sharkey and Sharkey 2010). This may depend on how robots are deployed: Wada and Shibata's research, which we saw in the previous chapter, suggests that appropriate use of robots can actually increase human connections. So, these different concerns can be mitigated in different ways, by technological fixes, policy changes, and responsible exercise of individual choice. But more broadly, this seems to be a particularly pointed example of the widespread if somewhat inchoate intuition that it is problematic to use robots to provide for social needs, introduced in the previous chapter. At least, it seems problematic given that our current crop of robots seems pretty far from the sort of thing one could plausibly consider a genuine person. While there may remain the possibility that a robotic agent such as Data from *Star Trek: The Next Generation* (Roddenberry 1987) could be a satisfactory companion for the elderly, today's robots, and those we can expect in the near future, are not plausibly considered to be people in any rich or interesting sense.

Recall that some object to sociable robots on the grounds that they "enchant" users. This is a concern for lucid as well as confused geriatric patients (for instance, in Turkle 2011; Grodzinsky, Miller, and Wolf 2015; Whitby 2011). However, this discounts the testimony of seniors such as those quoted earlier who, apparently lucidly, report benefits of interaction with robots. Respect for patients' judgments about what constitutes their well-being ought to be part of responsible geriatric care. And again, as earlier, the activation of emotions and social responses does not seem in itself problematic. Tearjerkers, comedies, and games in which, like Monopoly, people pretend to deal with valuables all seem like important resources in enriching patients' lives. The fact that robots stimulate our social and emotional responses does not, on its own, seem sufficient to make them ethically suspect.

And as before, sometimes when enchantment is harmful, blame may fall to the enchanted individual. The fact that patients sometimes misjudge their own social needs is not a good reason to prohibit use of sociable robots. Seniors (as indeed adults of all ages) sometimes make ill-advised choices when it comes to friends and companions; this does not necessarily license others to run their social lives. Even if some people are inclined to (by hypothesis) wrongly select robots as their primary companions, further argument is required to reach the conclusion that the robots are to blame. The picture is somewhat complicated by the fact that seniors may be particularly vulnerable to exploitative people who present themselves as companions, and it may be appropriate to protect them from the further harms made possible by the appearance of friendship. But enchantment alone still seems an unsatisfactory explanation of the badness of sociable robots, and a poor reason to limit their inclusion in a caregiver's toolkit. That it would be wrong to use a robotic companion to get access to a senior's financial resources does not mean that robotic companions are wrong, simpliciter.

"Substitution" accounts which hold that substituting robots for human interaction is bad, also do not fare well in this application (Sharkey and Sharkey 2010; Sorell and Draper 2014). Vallor's suggestion that appropriate use of social robots can facilitate human/human interaction seems to be borne out here. Introducing robot to geriatric facilities can actually increase both the quality and quantity of social interactions among patients, as we saw in the previous chapter (Wada and Shibata 2006).

Perhaps in an ideal world, patients and human caregivers would always find their interactions mutually rewarding, but the actual toll of caregiving work is significant, and if lifted may provide better on-the-ground results than hoping for human beings to act perfectly. If one reason for seniors' isolation is the difficulty associated with caretaking, and robots are used for stressful parts of eldercare, people may be more likely to prioritize spending time with elderly friends. Patients who are less troubled and more stimulated may be more pleasant and less tiring to interact with, making them more likely to be invited to participate in social events. This would not justify the initial neglect, but would tend to provide a practical work-around for an established problem. Furthermore, not every robot-for-human-caregiver substitution is bad. Vallor's point that robots may replace abusive caregivers or even just busybodies seems salient here (Vallor 2011).

The "deceptiveness" account of the badness of sociable robots may seem to be the most pertinent, because geriatric patients can be especially vulnerable to both loneliness and cognitive confusion of various kinds: a potent combination when it comes to deception about social connection. The picture may be complicated because benign deceptions may seem more permissible in geriatric care than in the general population. Consider, for example, some nursing homes' practice of installing fake bus stops near their facilities. Patients at high risk of wandering may encounter the bus stop, recognize the familiar sign, and wait for a bus until caregivers can come around to collect them (and often peaceably, if the patient has forgotten the original motivation for leaving). Both the fact that some paternalistic benevolent deception may be justified for cognitively compromised patients, and that it may be appropriate when it facilitates ease of use (as for example when it helps seniors to successfully interact with complex emerging technologies such as medical robots) means that potential deceptiveness is insufficient as a reason to restrict use. Instead, we need a tool that will help us distinguish good from bad deception.

Because of this, and because there are, as noted earlier in the chapter, powerful health benefits associated with using sociable robots in geriatric care, we need a conceptual tool that will help us to better evaluate the moral tradeoffs involved in any given use of the technology. The counterfeit account can do this. It says that the fact that sociable robots can provide patients with the subjective experience of friendship is insufficient for calling it good. This does not mean that subjective experiences are necessarily bad, all things considered, either. But it does mean that in giving geriatric patients robotic companions, we are not giving them friendship, so the good of friendship cannot be used to count potential deception as benevolent.

4. Ethical Implications for Geriatric Care

In the previous chapter, I invited the reader to imagine being given a choice between two lives. It seemed that that most would prefer the Genuine over the Truman Show life (Weir 1998), which showed that we value more than appearances in friendship.

What social robots do is give people the appearance of friendship. And the appearance of friendship can itself offer substantial instrumental benefits. Some of these benefits are available even when patients are aware that they are presented with a mere appearance, as with the senior who testified that interaction with the robot "healed my mind". But not every patient is in a position to realize this, even without active attempts to deceive. Nevertheless, when dealing with cognitively compromised or otherwise vulnerable individuals, responsibility seems to fall on the caregiver who presents a potentially deceptive object to the patient, rather than the patient. Well-intentioned or thoughtless counterfeiting is still counterfeiting. The question is, what ethical implications does this have, and how should this inform the decisions of caregivers, facilities using such robots, and social robotics designers?

Because friendship is not merely experiential, but our best evidence for it is grounded in our experiences, patients are at risk of being deceived about the nature of their relationship with robotic companions, and more so both because of cognitive impairments and social circumstances. The analogy developed in the previous chapter, between false friends and false coinage, can clarify of what exactly this deception consists. It will not answer a related question about when paternalism in the use of sociable robots is justified. But by getting clear on the nature of the deception, one can combine this with one's preferred theory of appropriate paternalism in geriatric care to decide when the benefits are worth the cost.

We saw that what people value about friendship is belonging to an emotionally interdependent relationship grounded in mutual caring. Robots are incapable of returning emotional dependence or attachment. So, following Aristotle, we should think that fooling people into thinking they have friendship, of the sort they value, by using robots that give the appearances of friendship, is analogous to passing off counterfeit coinage. And, as we also saw, the fact that seniors may never realize the deception, nor suffer tangible harms for their mistake, does not negate the wrong. Furthermore, as creatures who ourselves value friendship, it is bad for us to engage in counterfeiting friendship, even for benevolent purposes. Friendship, as a valuable component of human good lives, ought to be treated with appropriate consideration. Real interdependence matters. Earlier in this volume I have made some attempts to explore *why* this is so by examining the nature of friendship, but even if one is not satisfied with such explanations, it can still be taken to be a brute fact *that* it matters, and that providing people with the appearances of friendship is not the same thing as providing them with friendship, even when they can realize external goods associated with friendship thereby.

At the same time, some patients may simply be so compromised that they are incapable of grasping the appearance/reality distinction, with friends or elsewhere. In such cases, it may be more appropriate to say that for such people, the good of friendship is simply inaccessible to them. What remains, then, are considerations of goods other than friendship. And here, the external benefits of the appearance of friendship may carry the day.

5. Extending or Pretending?

To correctly apply this analogy, we should distinguish objects that falsely imply connections from those that facilitate them. Checks and credit lines are not currency, but they can extend economic relationships beyond those feasible in a cash economy. Telephones allow people to interact regardless of physical proximity. Other uses of technology might also constitute extension of real relationships. For example, a daughter of a man with dementia recorded herself reassuring her father and telling him to go back to bed. The recording was connected to a motion sensor in the man's home, and would play whenever he wandered the house in the middle of the night (Sharkey and Sharkey 2012). This might plausibly be viewed as a relationship extension that allows her to sleep while still reassuring her father, rather than a phony impression of a non-existent relationship.

Perhaps robots can be similarly used to reassure confused patients of their real social connections. In addition to telepresence robots, a sociable robot might be programmed to play a recording of a family member, for example, to help sooth a distressed patient during times when the family member cannot be physically present. Such a use could constitute an extension of a real relationship between people rather than a replacement of a human by a robot, and so would, on my account, be less problematic than a *de nuevo* robot companion.

But one might wonder how far this can be pressed: could *any* use of a sociable robot be justified by a sufficiently caring designer, manufacturer, or employee? Can care be transmitted from designer to patient in something like the same way that telephones and recordings help transmit conversants' care for each other? Suppose Dee the Designer cares about seniors. It might seem that anything Dee designs extends Dee's actual concern, so anyone who befriends a robot Dee designed is actually engaged in an extended friendship with Dee.

But this strategy fails. Such reasoning will not provide a blanket justification for using sociable robots in settings where caregivers' or designers' care is for patients generically, rather than particular patients with whom they enjoy close relationships. Friendships are highly particular relationships. Friends cannot be substituted for one another, and new friends cannot be swapped into existing friendships. So personal relationship extension will not occur on the basis of generic design features.

Money, by contrast, is highly fungible. Any instance of the generic kind "U.S. dollar bill" can be freely substituted for any other. But the very fact that it is intersubstitutable can be used to illustrate the problem with fungibility in friendships and other personal relationships. Gifts of money often seem inappropriate in personal relationships, precisely because money is generic. Contrast this with handmade gifts: even if ugly or otherwise lacking in fungible value, they often seem valuable because they reflect unique connection to a particular other person. So a generic concern for seniors will not make sociable robots associated with such concern genuine friends.

6. The Importance of Context-Sensitivity and Balancing Competing Goods

Application of this analogy between false friends and false coinage yields results for both designers of sociable robots, and caregivers who use them to provide care for geriatric patients. Designers should minimize confusion—and be clear about target audiences, as this will vary depending on the cognitive capacity of the patient. Caregivers need to consider specific needs of particular patients to provide instrumental benefits without causing confusion analogous to counterfeit currency, insofar as that is possible. Sometimes this can occur simply by explaining the nature of the object in question. Sometimes, a cartoonish or otherworldly appearance can help—Paro the robot seal may be less deceptive than a robot resembling a child, for many patients.

It also cautions us to be moderate and sensitive to context in our assessments of badness. Not every use of sociable robots constitutes an analogy to counterfeiting, and not every social good requires genuine friendship. We should recognize that the appearance of friendship can "toy with" our emotions without being morally bad (compare to the "enchantment" of films and stories). As discussed earlier, the subjective experience of friendship provides other goods, such as physical health and alleviation of perceived loneliness. Sometimes, competing goods will need to be balanced against each other. But clarifying which goods are involved, and how competition can be minimized, will allow us to make better decisions and reduce ethical costs.

7. Conclusion

Aristotle's analogy directs us to consider whether people are fooled into thinking they belong to groups, and who is to blame for the deception. It thus provides guidance on the ethical use of sociable robots. As cognitive confusion increases, the risk of deception goes up. And while the deception may be justified by the tangible benefits of alleviating subjective loneliness, it cannot be successfully claimed that robots are satisfactory sources of friendship, an important element in patients' quality of life.

The powerful benefits and seemingly intractable costs of sociable robots identified lead me to believe that we should not expect a simple policy decision on whether or not their use is ethical. Rather, their implementation in eldercare involves a range of thorny issues. I think that in such matters, we should not look for a uniform principle to guide us. Details of context matter, and identifying relevant features of context is a nontrivial task. Most real-life contexts in which we must make ethical decisions involve tradeoffs between competing goods (e.g., between physical health and autonomy), and it is no different for decisions about deploying sociable robots. Complicating matters further, the varying cognitive, emotional, and volitional abilities of geriatric patients may influence what counts as a good choice for a given subject.

So what, then, should we look for, if not a uniform principle that tells us what to do? Ideally, we want to make wise and careful decisions that are appropriately sensitive to context, but we may need help to get there. An analogy like that developed in the previous chapter can serve as a roadmap to help take stock of the territory, even if it cannot tell us exactly where to go. A good tool can help people think through particular cases and identify important costs and benefits in order to make intelligent tradeoffs.

The issues surveyed so far suggest that an appropriate tool for navigating ethical issues will incorporate the following considerations. As we have seen, some patients recognize that they are interacting with robots, while others, owing to various kinds of cognitive disorders, do not. This is the difference between enchantment and deception. In enchantment, one recognizes that the thing with which one is interacting is not a person, but because features of the object appeal to one's emotional patterns of response (such as big cute "eyes", high-pitched vocalization, behavior modification in response to voice, touch, or facial expression), one goes ahead and interacts as if the item were a person. In robust deception, these or other features are sufficient to convince the individual that the object is a person (as with the dementia patients who thought AIBO was a baby). In some cases, deception might be unproblematic, because the results are benign: if, for example, a user takes a file folder in a GUI to be an actual file folder, no harm seems to be done. But with friendship, it (constitutively) matters that one's friends reciprocate, and so deceiving a person about the nature of the thing taken to be a friend is a salient ethical consideration.

In evaluating the well-being of patients, a broad and rich conception of well-being should be utilized. It should not be limited to thin, easily-operationalized concepts like subjective pleasure or lowered incidence of dementia, but should also include whether these patients are living good lives of the sort we consider choice worthy; whether their desires are really being met or only appear to be satisfied. Instead, something more eudaimonist, which includes both physical and psychological health, as well as richer and less easily quantified goods like friendship, would be desirable.

In addition to the well-being of patients, the impact of sociable robots on caregivers, designers, and patients' families and friends should be taken into consideration (Vallor 2011). As noted earlier, using robots in eldercare may help reduce the disproportionate burden of care work on some family members. But there are also costs. If substituting robot interactions for human ones tends to weaken connections to living friends, that could also affect people beyond the patient. Or if it constitutes treating the elderly as less valuable and deserving of fewer goods than the rest of us (as, for example, taking geriatric patients to only require the subjective experience of friendship, while the rest of us crave the genuine article), that would be a reason to consider it ethically bad.

Especially, if using robots to satisfy elderly people's social needs causes us to devalue friendship, this seems to count as a reason against their use even if the patients themselves do not feel the impact of this effect. Friendship is a valuable human institution, and thriving as human beings may require us to take it seriously and treat it as something about which it is bad to deceive people. And if using robots to deceive patients into thinking they have friends when they do not makes us more comfortable with deception about who a person's friends are, we have reason to tread with caution in their design and implementation.

Part of the challenge here has figuring out how to balance respect for patients' values, including the value of friendship, that have been established over many years, with health considerations that can be advanced by presenting them with the appearance of friendship, and factoring in the potential for confusion when appearances that have long been associated with one kind of phenomenon—human relationship—are now caused by a very different thing. In the following chapter, ethically salient features of context will be somewhat different. The big-picture concerns about how to appropriately make decisions that protect the well-being and interest of patients will be the same. But the application will involve teaching young children in ways that will help establish values and expectations, rather than work with pre-existing ones. In consequence, the analysis will be somewhat different, even though some concerns will be shared.

Bibliography

Banks, Marian R., Lisa M. Willoughby, and William A. Banks. 2008. "Animal-Assisted Therapy and Loneliness in Nursing Homes: Use of Robotic Versus Living Dogs." *Journal of the American Medical Directors Association* 9 (3): 173–77.

Borenstein, Jason, and Yvette Pearson. 2010. "Robot Caregivers: Harbingers of Expanded Freedom for All?" *Ethics and Information Technology* 12 (3): 277–88. doi:10.1007/s10676-010-9236-4.

Fratiglioni, Laura, Wang Hui-Xin, Kjerstin Ericsson, Margaret Maytan, and Bengt Winblad. 2000. "Influence of Social Network on Occurrence of Dementia: A Community-Based Longitudinal Study." *The Lancet* 355 (9212): 1315.

Grodzinsky, Frances S., Keith W. Miller, and Marty J. Wolf. 2015. "Developing Automated Deceptions and the Impact on Trust." *Philosophy & Technology* 28 (1): 91–105. doi:10.1007/s13347-014-0158-7.

Kanamori, Masao, Mizue Suzuki, and Misao Tanaka. 2002. "Maintenance and Improvement of Quality of Life Among Elderly Patients Using a Pet-Type Robot." *Nihon Ronen Igakkai Zasshi. Japanese Journal of Geriatrics* 39 (2): 214–18.

Parks, Jennifer A. 2010. "Lifting the Burden of Women's Care Work: Should Robots Replace the 'Human Touch'?" *Hypatia* 25 (1): 100–20. doi:10.1111/j.1527–2001.2009.01086.x.

Ponte, Emily. 2014. "Recent FDA Medical Device Regulation and Its Relevance to Robotics." *Tech Policy Lab*. Accessed January 27. http://techpolicylab.org/fda-medical-device-regulation/.

Roddenberry, Gene. 1987. "Star Trek: The Next Generation." NTSC. *Episodes 1–176*. Paramount Television.

Sharkey, Amanda. 2014. "Robots and Human Dignity: A Consideration of the Effects of Robot Care on the Dignity of Older People." *Ethics and Information Technology* 16 (1): 63–75. doi:10.1007/s10676-014-9338-5.

Sharkey, Amanda, and Noel Sharkey. 2010. "Living with Robots: Ethical Trade-offs in Eldercare." In *Close Engagements with Artificial Companions: Key Social, Psychological, Ethical and Design Issues*, edited by Y. Wilks, 245–56. Natural Language Processing. Amsterdam: John Benjamins Publishing Company. https://books.google.com/books?id=EPznZHeG89cC.

———. 2012. "Granny and the Robots: Ethical Issues in Robot Care for the Elderly." *Ethics and Information Technology* 14 (1): 27–40. doi:10.1007/s10676-010-9234-6.

Sorell, Tom, and Heather Draper. 2014. "Robot Carers, Ethics, and Older People." *Ethics and Information Technology* 16 (3): 183–95. doi:10.1007/s10676-014-9344-7.

Sparrow, Robert, and Linda Sparrow. 2006. "In the Hands of Machines? The Future of Aged Care." *Minds and Machines* 16 (2): 141–61. doi:10.1007/s11023-006-9030-6.

Sullins, John P. 2008. "Friends by Design: A Design Philosophy for Personal Robotics Technology." In *Philosophy and Design*, 143–57. The Netherlands: Springer. http://link.springer.com/chapter/10.1007/978-1-4020-6591-0_11.

Tamura, Toshiyo, Satomi Yonemitsu, Akiko Itoh, Daisuke Oikawa, Akiko Kawakami, Yuji Higashi, Toshiro Fujimoto, and Kazuki Nakajima. 2004. "Is an Entertainment Robot Useful in the Care of Elderly People With Severe Dementia?" *The Journals of Gerontology: Series A* 59 (1): M83–85. doi:10.1093/gerona/59.1.M83.

Turkle, Sherry. 2011. *Alone Together: Why We Expect More from Technology and Less from Each Other*. New York, NY: Basic Books. https://books.google.com/books?id=J2ine5sIlkgC.

Vallor, Shannon. 2011. "Carebots and Caregivers: Sustaining the Ethical Ideal of Care in the Twenty-First Century." *Philosophy & Technology* 24 (3): 251. doi:10.1007/s13347-011-0015-x.

Wada, Kazuyoshi, and Takanori Shibata. 2006. "Robot Therapy in a Care House—Its Sociopsychological and Physiological Effects on the Residents." In *Proceedings 2006 IEEE International Conference on Robotics and Automation, 2006. ICRA 2006*, 3966–71. doi:10.1109/ROBOT.2006.1642310.

Weir, Peter. 1998. *The Truman Show*. 35 mm. Paramount Pictures.

Whitby, Blay. 2011. "Do You Want a Robot Lover?" In *Robot Ethics: The Ethical and Social Implications of Robotics*, edited by P. Lin, K. Abney, and G.A. Bekey, 233–49. MIT Press. https://books.google.com/books?id=oBb-lt3l4oYC.

Wilson, Robert S., Kristin R. Krueger, Steven E. Arnold, Julie A. Schneider, Jeremiah F. Kelly, Lisa L. Barnes, Yuxiao Tang, and David A. Bennett. 2007. "Loneliness and Risk of Alzheimer Disease." *Archives of General Psychiatry* 64 (2): 234–40.

6 Counterfeit Currency Versus Monopoly Money

Using Appearances to Build Capacities

1. Introduction

In this chapter, I argue that in order to properly equip children with autism spectrum disorder to make good choices about social connection, therapeutically beneficial robot "friends" need to be designed and deployed so as to support their ability to distinguish between real and apparent friends. This means being careful with attachments and appearances of robots, and not trading therapeutic effectiveness on one front for broader concerns about the well-being of these children, taken so as to include their ability to access the human good of friendship. At the same time, we should be careful to give them the freedom to choose once they are sufficiently equipped to do so, a topic that will be revisited in the next chapter.

Part of the problem here, as in the previous chapter, is that the designers of these technologies are aiming at users' needs only indirectly; they are mediated by caregivers and therapists who are in relatively paternalistic positions. The geriatric care worker, home caregiver, parent, or child therapist must, to some degree, evaluate for the patient (whether old or young) what constitutes satisfaction of the patient's social needs. Whether by reason of frailty, mobility restrictions, senile dementia, or childhood social developmental disorder, the patient is by hypothesis incapable of independently procuring social goods, and while they can and should be consulted, their interests must of necessity be moderated by those who are in a position to both care for and evaluate their best interests. This makes a working theory of the social good of friendship helpful and important, as a means to help paternalistic caregivers to clearly and objectively evaluate what is good for the patients, untainted by considerations of convenience or (mistaken) patients' apparent contentedness with subpar care.

In this case, the details of care vary considerably from those of the previous chapter. Rather than dealing with increasing social isolation and declining faculties of a geriatric population, caregivers are in the position of equipping often very young children with skills and resources to develop their first fulfilling genuine social connections (or at least those outside the birth family). Autism spectrum disorders (hereafter ASD) are characterized by early childhood onset impairments of social function, from linguistic disorders, to difficulty recognizing facial and bodily expressions, to problems engaging in eye gaze, imitation, cooperative activities, turn-taking, and joint attention (a condition which, in its fullest expression, involves two or more individuals jointly attending to an object of attention, where each is aware of the other's attention and the fact that each is aware of that same object) (American Psychiatric Association 2013). While autism spectrum disorders cannot be cured, early intervention can help children develop these skills to a greater degree. Practice recognizing facial expression, for instance, can help ASD children to develop this skill, which will in turn help them to function socially, and to enjoy social goods that might otherwise be inaccessible to them.

2. Socially Assistive Robots for ASD Therapies

Technologies that help ASD patients to practice these skills are therefore valuable. And the fact that many children with ASD are fascinated by technology makes this a promising avenue for those interested in therapeutic interventions. Techniques that capture and hold the interest of young patients, especially those whose disorders make directing and maintaining attention quite difficult, are of great interest. And evidence suggests that robots have much to offer on this front. Some research (Dautenhahn and Werry 2004; Robins, Dautenhahn, and Dubowski 2006) has found that when presented with the choice, children with ASD tend to prefer interacting with robots to interacting with human companions. Other studies (such as Duquette, Michaud, and Mercier 2008) found that patients with ASD are more responsive to facial expressions such as smiles and focus for longer when interacting with robots versus humans.

Although most studies of these technologies are still exploratory, early results are promising on a number of fronts. Researchers are looking at ways to use robots for practicing things such as body awareness (Costa et al., 2013), asking questions (Huskens et al., 2013), imitating a confederate (Greczek et al., 2014), collaborative play (Wainer et al., 2014), joint attention (Anzalone et al., 2014), and tactile play (Robins and Dautenhahn 2014).

Robots such as *Probo*, billed as the "huggable robot friend", are designed specifically for ASD therapy. Other programs recruit existing commercial robots like NAO or Keepon for therapeutic purposes, or build custom robots for specific studies. Various robots can be used for training in everything from facial or body posture recognition, to physical contact, to imitation, to cooperative game play, to directing gaze or joint attention exercises (separately or together). So it might be best to think that different features are implicated in different abilities. And ASDs themselves are called *spectrum* disorders for a reason: different children can present with very different degrees and combinations of symptoms. Some are quite high-functioning, while others have severe difficulties with language and most forms of human contact.

In a survey of research on this topic, Scassellati et al. conclude that socially assistive robots are highly promising for ASD therapies. But, they note, it is unclear what is doing the work here. They do offer several hypotheses:

> Perhaps the simplified social cues that robots present result in less overstimulation of the children; perhaps robots offer more predictable and reliable responses than those from a human partner with ever-changing social needs; perhaps robots trigger social responses without the learned negative associations that some children have with human-human interactions; and perhaps the exaggerated social prompts that robots provide are better triggers for social behavior than the nuanced and subtle social prompts from a human partner.
> (Scassellati et al., 2012, 292)

The answer might be any or all of these (a perfect storm), or it might vary from context to context or child to child. The complexity and robustness of the phenomenon make it difficult to navigate. But because it involves activating and exercising children's social responses, it seems important to consider the ethical implications of using such technologies.

Some guidelines for practice seem relatively straightforward. For example, if Scassellati et al. are correct about the reasons for ASD patients' preference for robots over humans, it will be important for therapies to take these factors into account. Less-anthropomorphic robots may initially be more appealing to patients than highly-realistic ones (or human therapists), and therefore make good tools for introducing children to exercises. Increasingly complex modifications or models of robot could be used to match tasks to the patient's skill level as the patient progresses. And some studies (Goodrich et al., 2011; Wainer et al., 2014) are exploring the effects of alternating robot-assisted practice sessions with similar scenarios involving human confederates. This may help children to transfer their abilities to naturalistic human-human interactions.

3. Concerns About Using Robots With ASD Patients

The promise of such technologies is apparent to many people concerned about ASD. In a survey on therapeutic use of robots in treating ASD, 85 percent of respondents thought it was ethically acceptable to use these technologies for this purpose, among a population where 60 percent of the general population is skeptical about the morality of using robots for caregiving. This may not be representative of the general population's views. The respondents surveyed were not a random selection, but drawn from parents, therapists, and advocates for ASD patients. But this makes the results more interesting, not less. As stakeholders in the treatment of ASD, their receptivity to this technology is noteworthy. But another interesting result from the survey is a range of concerns about unethical uses of this technology, even given widespread acceptance of their general potential in this application. For example, a majority of respondents thought it was unethical to use these robots in such a way that children believed these robots were their friends (only 43 percent thought it would be ethical to do so). Far more thought it permissible to use robots that were zoomorphic (74 percent) than anthropomorphic (55 percent). And far more were comfortable with the use of telepresence robots operated by a therapist than autonomous robots (Coeckelbergh et al., 2016, 58).

It is possible, of course, that these simply reflect uninteresting suspicions or biases on the part of users. The cynical designer might take them to heart only insofar as they give information about the sorts of accommodations most likely to make relevant user bases adopt and utilize the technology, much as a manufacturer of a medicine might take account of customer preferences for taste only insofar as it helps the manufacturer decide how to flavor the concoction. But I think there is something deeper at work here. ASD, as a cluster of disorders, is difficult precisely because sociality is so important to us. And as we have seen elsewhere in this volume, while we have strong feelings and intuitions about friendships and other interpersonal relationships, there has been relatively little theoretical work done to clarify the objects of these feelings and intuitions, at least in value theory. It would benefit us to spend some time considering what exactly we are aiming at when we attempt to assist with disorders that impair social functioning, and to consider carefully how well the tools for the job can help us to achieve not just particular behavioral markers, but the rich human goods of friendship. The fact that even stakeholders who are open to the therapeutic benefits of socially assistive robotics are concerned, specifically, about robots that present the appearance of friendship, especially autonomous ones (those that are not mere bridges that ultimately connect human beings), and concerned about children believing that such robots are their friends, seems to me to be not just a concern about new technology or a fear of the unknown, but a concern about friendship itself, and how well our attempts to address particular social skills end up supporting versus undermining it.

One possibility to be considered is that the risk involved in presenting children with robot "friends" is the hazard, generally, of attachment. As Coeckelbergh et al. (2016) note, it is difficult to use socially assistive robotics in this context without running at least some risk that the children will become attached to these robots. And to some degree, that can be therapeutically beneficial: attachment to these objects can help the children to practice, to persist, and to find exercise of their growing skill set rewarding. Of course, this also makes them vulnerable to loss. If at the end of a treatment program an expensive robot is reclaimed by the healthcare provider, this could seem to the child as devastating as the loss of a human friend, if the child were sufficiently attached. Even in the case of cheaper or individually-owned robots, the breakage or loss of the item could be cause for grief. In this respect, robots might be like beloved pets, or even relatives: loss of these objects of attachment might constitute the children's first experience of grief. This concern is echoed in Riek and Howard's code of ethics for human-robot interactions (Riek and Howard 2014).

But this would be an odd reason to think it unethical to represent robots as friends. After all, many people think it is good for children to have close attachments to companion animals or elderly relatives, despite the fact that they will almost certainly lose these attachments eventually. And while children may become attached to therapists, this does not seem to be a reason to avoid therapy with a determinate end date. And, of course, some of these problems come with practical work-arounds: children who become attached to robots might be provided with their own "friend" to keep at the end of the therapy, even if such costs need to be factored in, or additional efforts could be made to repair or keep home therapeutic robots in service for the psychological well-being of the child (assuming that grief at its loss represented a threat to well-being—itself a contentious claim).

In short, taking attachment to be the problem because attachment opens the child up to the possibility of loss is not consistent with how we normally think about the value of attachment. We do not shield our children from forming attachments to friends, because we or they might move, or discourage them from becoming attached to companion animals with short lifespans or elderly friends or relatives nearing the end of their expected lifespans. This does not rule out the possibility that the users polled fail to think clearly about the issue. But it does suggest that, if we are to treat them as rational valuers with at least a mostly-coherent set of values, we might do well to consider other explanations for their response patterns.

Grodzinsky, Miller, and Wolf (2015) raise concerns about introducing vulnerabilities in human-robot interactions by encouraging people to trust artificial agents. If one thinks of a robot as a friend, this will presumably make one disposed to trust the robot in the ways appropriate to a friend. But because artificial agents can be hacked, or duplicates introduced that are indistinguishable to the user from the original, malicious individuals can exploit this trust to get at secure information. This surely makes good security a concern for manufacturers of artificial agents that may engender such trust in users, both making devices personalized in ways that make duplication difficult, and protecting them against malicious hacking. But the same risk is present wherever a person trusts an artifact, from an ATM to a bank website. It is a reason to be responsible, not to avoid the technology altogether. This does not seem sufficient to explain why we should be resistant to technology that facilitates in children the belief that these robots are their friends.

4. Putting the False Friends and False Coinage Analogy to Work

The false friends/false coinage analogy introduced in Chapter 4, by contrast, neatly organizes and explains these response patterns among ASD stakeholders. They value friendship and are open to the possibility that socially assistive robots can help these children to foster friendships of their own. But they do not want children to think that what these robots offer *is* friendship. Nor do they want these robots to accidentally mislead children into believing this is so (this would explain the widespread preference for zoomorphic over anthropomorphic robots).

The last data point is perhaps the least obvious, but I think it coheres well with the friendships/economies analogy. ASD stakeholders widely preferred telepresence robots to autonomous ones. Autonomous robots offer the advantages of convenience and predictability. But telepresence robots connect children to other people, albeit by means that these children find less intimidating and more appealing than face-to-face human interactions (perhaps for the reasons suggested by Scassellati et al. 2012).

There could, of course, be other explanations for this last preference. Perhaps they are worried about safety (although safety concerns are among the top priorities of designers of such products, and introducing human error into the equation hardly seems likely to be safer). Perhaps they believe human operators can be more observant, flexible, and adaptable, providing a higher quality of customized therapy to children than an automated robot operating according to a strictly coded pattern of behavior. This is a concern some roboticists are working to address (Bekele et al., 2013) but is in principle surmountable with appropriate technological innovation. Or perhaps they think that human interactions will be more *challenging* to children, and so help them to push their boundaries and develop appropriate skillsets, rather than stay within their comfort zones, even if this to some extent undercuts the appeal of these robots to patients. In both cases, these seem to amount to concerns about the quality of therapy available via telepresence robots versus autonomous ones.

A story from an exploratory study investigating the use of a socially assistive robot to teach ASD children various elements of collaborative game play illustrates this concern. The researchers found that the children were more interested in and engaged with the robotic companion than the human one. But data from the results seemed to suggest that they actually performed better at their assigned tasks with the human partner. The reason for this had to do with the predictability of the robot, KASPAR:

> because the robot was programmed to prompt a child to choose a shape if they remained inactive for 5 consecutive seconds, as well as to repeatedly attempt to take the initiative in choosing shapes when the child did not respond to two consecutive prompts, one child discovered that KASPAR would essentially speak to them every 5 seconds provided that they did not play at all. This allowed the child to stare raptly at the robot for long periods of time without responding to the robot's prompts, until the child's carer jogged him out of this routine.
>
> (Wainer et al., 2014, 60)

Being an exploratory study, the number of participants was small enough that this threw off the results. The researchers recommended varying the robot's behavior and study design so as to avoid this problem in future research. But it illustrates the problem: preprogrammed behaviors cannot be expected to anticipate and respond to children's individual needs, even if they are highly engaging to the users.

But there is an advantage to explanatory unity. If a cluster of responses can all be explained by one theory, rather than requiring a different one-off explanation for each, that has the explanatory advantage of simplicity. And this account does so. People do not seem to want ASD children to have robot friends instead of human ones, even if the children at least initially prefer the robot ones. And even if other factors also contribute to some of the responses, it is noteworthy that this account of friendship so neatly matches so many concerns.

5. Using Pretense for Instruction

It is, then, worth investigating what happens when we turn to this analogy, not just to explain what could go wrong, but to see how it can help us to make and use such robots wisely and well: not to get caught up in overly simplistic questions like "autism spectrum disorder therapy robots: bad or good?", but to think about what they are for, how they work, and what constitutes wise versus foolish use of them.

In thinking about this issue, it is helpful to turn to the money analogy for a moment. Economies, as I said earlier, are interdependent social groups connected by a currency. Learning to function well as an economic agent requires learning how to use money, and how to use it well. Few if any people take to this naturally; in this respect, perhaps, we are all developmentally incapacitated with respect to money. But a variety of tools can help us learn how it works, how to handle it, and how to use it well.

Consider the game Monopoly. In this game, children are given brightly colored paper slips printed so as to (somewhat) resemble money. According to the rules of the game, players start with a certain amount of "money", but by trading it for goods that offer one the prospect of collecting more of it from others, investing it, and giving it up as an in-game penalty, children learn how money works (albeit in a simplified and cartoonish format) and practice using it themselves. Because money matters, this and similar games can be good for children's development. Playing with toy money in a controlled setting can help one to experiment, learn, and develop skills relevant to participating in real-world economies. And making it fun in the process can increase the time and consistency of this learning. At the same time, it matters that the "money" only loosely resembles real money. It is important that the children (and adults!) be able to distinguish the genuine article from enjoyable simulacra. Furthermore, the children should be able to understand and appreciate the difference between the value of real versus toy money.

6. Objection From the Subjective Nature of Value in Friendship

At this point, the skeptical reader may object. The disanalogies between friendship and money might seem too important to accept what I have said above. After all, money as a social institution is highly formalized, and learning the "rules" matters, while friendship is highly fluid, and there are few consistent principles to be found. In fact, what few general principles there are seem to be so abstract as to be useless when it comes to practical matters. (As we saw in Chapter 2, one anthropologist went so far as to call friendship an "institutionalized non-institution" (Paine 1969, 514)). The value of money is capable of being precisely calculated, relative to a wide variety of other goods, including other currencies. But friendship's value is highly difficult to pin down. And money's value is by definition intersubjectively valuable, while what makes someone a good friend, or a friendship a successful one, seems highly dependent on the particular quirks of the individuals. So long as involved parties seem happy and no one seems worse off for the association (and even that is controversial, as we saw in Chapter 3), one might think, outsiders have no grounds to justify criticizing the relationship. This is a thought that might be modified somewhat based on results I have reached in Chapter 3, but even this is restricted to concerns about friends who make each other worse off.

Because of this cluster of concerns, it might seem inappropriate to use the model I have suggested to capture ethical concerns about socially assistive robotics in autism spectrum disorder therapy. After all, if what makes someone a good friend is subjective, then there is nothing better or worse about patients' befriending robots than about anyone else's befriending a human being. Just because I would not be happy in such a relationship, does not mean you cannot be. (This can be a common experience, upon meeting an acquaintance's close friend: "they seemed very nice but not my cup of tea".) And if the rules for friendship are fluid and to a large extent fixed by the preferences of the friends, then there is no point in training children in some particular skill set, because the friendship should be adapted to the needs of the individuals in any case, rather than the other way around—and in many ways robots can seem much better suited to satisfy ASD children's social needs than human beings, for the reasons noted by Scassellati et al. (2012).

This would not mean that robot therapy should be discontinued. Rather, it would indicate that the skills children acquire by these means are largely of instrumental value: for helping them to secure employment, engage in business and economic transactions, participate politically, and otherwise enjoy the goods provided most conveniently by other people in our present social configuration. Insofar as these children may come to consider these robots their friends, this would be no cause for concern. It would only be a difference in preferences, not a mistake in values. The concerns of the respondents in Coeckelbergh et al.'s (2016) survey would, on such an account, have inadvertently fallen into the habit of projecting their preferences onto other's choices and finding them wanting, rather than identifying a matter of deep moral concern, when they think it would be bad for children to think robots are friends. We might do well to restrict the likelihood of psychological damage caused by loss of an object of attachment, at least under some circumstances. But we should not moralize the issue, assuming there is something inherently wrong with robot friends.

To some extent, I dodged this issue in the previous chapter, because geriatric patients have established values that presumably include the importance of real over merely apparent friendship. But in this case, the question is not whether this reflects (especially very young) children's established values, but how interacting with robots will shape their values. This seems like an example of Coeckelbergh's worries about using established value frameworks to assess emerging ones, encountered in Chapter 4.

Aristotle thought "no one would choose to live without friends even if he had all the other goods". But he might (understandably) have failed to take into account the rich variations found across humanity, especially in light of our new technical abilities. In the science fiction film *Blade Runner*, the character J. F. Sebastian is an eccentric inventor who lives in relative isolation. When he takes home Pris, an android who is grappling with the possibility that she may herself be a person who cares about her comrades, she expresses concern. "Must get lonely here, J. F." "Not really", he answers. "I MAKE friends. They're toys. My friends are toys. I make them. It's a hobby. I'm a genetic designer" (Scott 1982).

His toys are stylized, animated versions of traditional playthings: teddy bears, soldiers in uniform. They move and speak in a simplified, exaggerated, and cartoonish manner. While the "replicant" androids (of which Pris is one) are made to resemble humans as closely as possible, sometimes requiring complex tests to distinguish them, these would not fool anyone into thinking they are human. And even if they were difficult to distinguish, J.F. should be able to tell the difference—after all, he made them. And yet he seems happy to consider them his friends, and may even prefer them to human ones.

This, if correct, would suggest that ethical issues involving such robots would run in a different direction. Perhaps, for example, we should be asking ourselves how to incorporate concerns about virtue in friendship into robot design. How can a robot embody the virtues? How can a robot engage with a child so as to share a life that is worthwhile? How much should adult engineers, therapists, and parents monitor and tailor the robot to suit the child's particular needs, and how much should the child adapt? After all, friendships between humans involve give and take.

7. Response From the Capabilities Approach

But before tackling such questions, I think there is reason to take concerns about ASD patients thinking robots are their friends seriously, without simply imposing some people's conceptions of friendship on others.

To start with, it is uncontroversial that many people value friendship. The challenge here is that we do not yet know what many of the ASD patients in question will value, particularly when considering very young children. Perhaps they will value human friendship but find that, despite the best efforts of therapy, the advantages of robot interactions make them the more practical choice of companion. Perhaps therapeutic interventions will allow or assist them with the formation of human friendships. Perhaps they will find the human/robot distinction uninteresting and be indifferent between the two. Or perhaps they will find relationships with robots genuinely more fulfilling. The point is that we do not know. Given this uncertainty, how should we proceed?

The capability approaches advocated by Martha Nussbaum and Amartya Sen suggest a promising strategy (Nussbaum 2001; Sen 1989). The details of their respective accounts should not make a difference, here. Both are concerned to integrate objective standards for evaluating well-being across societies on an international scale, while remaining respectful of cultural differences and refraining from unreflectively imposing one's own moral framework wholesale onto others. One obvious approach, often favored by economists, is to look at how successfully people's desires are satisfied. This would allow for individuals to have very different preferences and desires, while measuring whether individuals are doing well by their own lights. But this approach runs the risk of mismeasurement of well-being, because it overlooks some problematic ways that individuals' desires can be shaped by culture, conditioning, and circumstance. In particular, it misses so-called adaptive preferences. Adaptive preferences are formed when individuals who are subjected to long-standing hardships adapt their preferences to fit within those limitations. For example, in communities where women's oppression is widespread and involves restricted access to resources like food and medicine, women may report that they have adequate nutrition and healthcare but, upon examination by a medical professional, be evaluated as suffering from malnutrition and untreated but treatable medical conditions. There seems to be a strong case for thinking that their reported satisfaction of preferences is not an accurate measure of their well-being.

To avoid both imperialistic imposition of researchers' values on other cultures, and measurements that miss the real psychological impact of long-standing hardship, the capability approaches broadly specify particular capacities that seem to be cross-culturally important, even if expressed differently in different cultures. For example, access to adequate nutrition seems to be a universal concern, even though different cultures may consume very different diets and in different ways (as for example through different meal structures), and some individuals, for political, moral, or religious reasons, may choose to engage in extended fasting. The capabilities approach holds that it is important that individuals be *capable* of obtaining adequate nutrition, but remains neutral with respect to the form in which it is available (for example, whether the carbohydrates take the form of rice, wheat bread, or yams) or whether any given individual chooses to exercise this capacity.

The evidence that friendship is an important good is fairly compelling. Across a wide range of contexts, times, and cultures, from Aristotle's Athens to the Buddhist philosophers of the Indian subcontinent to Kongzi's (Confucius's) China to contemporary America, friendship has been identified as a valuable element of a good life. While there are individuals who may choose to live without friends, in many cases this may look more like a tradeoff than a purely values-driven choice. And, most importantly, many individuals with even severe ASDs enjoy rich and rewarding friendships (see, for example, Temple Grandin's (2006) discussion of friendship in *Thinking in Pictures*). Even in the fictional universe of *Blade Runner*, J.R. Sebastian's choice to live with robot friends seems not entirely independent of the fact that he clearly feels awkward around human beings, among other things suffering from a rare disorder that causes accelerated aging. Perhaps he has ended up concluding that robots are the best companions for him because human society has not made room for him, not because they are inherently preferable, and the moral burden ought to fall on the rest of his community for so failing. This does not mean that individuals may not freely choose to live as hermits, or, alternatively, to strive to embody universal compassion without emphasizing particular ties. But it does mean that the widespread valuing of friendship calls out for an explanation, and the exceptions can often be integrated into a theory that it is a near-universal human good, on the order of something like adequate nutrition.

Given this, and the fact that interacting with others requires a particular skill set that ASDs impair, the case for equipping children with the requisite skill set so as to be capable of enjoying friendship becomes stronger. And here, the fact that friendship seems to constitutively require a distinction between real friends and merely-apparent ones becomes quite salient. This seems to be a lesson from Chapter 4. In order for people to enjoy the good of friendship, they must be capable of appreciating this distinction, even if they do not actively reflect on it. To enjoy merely the appearances without thinking the reality makes a difference just seems to be something else, and calling it friendship would be a category mistake. This does not mean that capable adults cannot choose robot companions, in addition to or instead of human friends. But it does mean that the phenomenon of friendship must not be misrepresented in the process of teaching children skills to facilitate it. Creating and deploying socially assistive robots that cause children to believe these artifacts are their friends, and not merely to enjoy pretending as if they were, becomes an important factor to take into consideration. Monopoly "money" can be used playfully to teach money skills but should not be represented as actual money to users, and appearances and contexts of use should support users' ability to distinguish between real and pretend currency. Once patients have the capacity to enjoy friendship and appreciate what it is, they will be better positioned to decide for themselves if this is a good worth pursuing. They will also, in the process, become better acquainted with the nature of robotic companions.

8. Conclusion

In this chapter, I have argued that we should take stakeholders' concerns seriously, especially worries about the potential for children with ASD to view socially assistive robots as friends. While it is tempting to interpret this concern as a matter of making the children vulnerable to exploitation of trust if malicious parties take control of the machines, or as introducing the risk of grief if the child should lose the "friend", I argued against these attempts to reduce the concern to extrinsic or contingent harms. Instead, I have made the case that the relationship itself is troubling if taken by the child to be friendship.

The fact that friendship is a valuable human good, however, also gives reason to make use of tools, including robots, which can help patients with ASD to increase their capability of enjoying the good. The fact that deception about friendship seems unethical does not mean that we should discard playful means of helping people to learn and develop relevant skills. Monopoly "money" and other toy currencies provide a helpful model for developing and using these tools are *not* real friends, and designing them to be zoomorphic or to resemble fictional creatures rather than realistic humanoids, seem like important steps toward this goal.

Doing so will help equip ASD patients to make informed choices about their future social engagements. In effect, they will be at least as well-positioned as the average consumer to decide whether to include robotic companions in their lives, in addition to or in place of human friends. At this point, then, the discussion moves on from paternalistic questions about how to help those ill-positioned to make and sustain their own social connections. We now turn to the ethics of mass market companionate robots.

Bibliography

American Psychiatric Association, ed. 2013. *Diagnostic and Statistical Manual of Mental Disorders, Fifth Edition*. Washington, D.C: American Psychiatric Association. http://dsm.psychiatryonline.org/doi/book/10.1176/appi.books.9780890425596.

Anzalone, Salvatore Maria, Elodie Tilmont, Sofiane Boucenna, Jean Xavier, Anne-Lise Jouen, Nicolas Bodeau, Koushik Maharatna, Mohamed Chetouani, David Cohen, and MICHELANGELO Study Group. 2014. "How Children with Autism Spectrum Disorder Behave and Explore the 4-Dimensional (Spatial 3D+ Time) Environment during a Joint Attention Induction Task with a Robot." *Research in Autism Spectrum Disorders* 8 (7): 814–26.

Bekele, Esubalew T., Uttama Lahiri, Amy R. Swanson, Julie A. Crittendon, Zachary E. Warren, and Nilanjan Sarkar. 2013. "A Step Towards Developing Adaptive Robot-Mediated Intervention Architecture (ARIA) for Children With Autism." *IEEE Transactions on Neural Systems and Rehabilitation Engineering* 21 (2): 289–99. doi:10.1109/TNSRE.2012.2230188.

Coeckelbergh, Mark, Cristina Pop, Ramona Simut, Andrea Peca, Sebastian Pintea, Daniel David, and Bram Vanderborght. 2016. "A Survey of Expectations About the Role of Robots in Robot-Assisted Therapy for Children with ASD: Ethical Acceptability, Trust, Sociability, Appearance, and Attachment." *Science and Engineering Ethics* 22 (1): 47–65. doi:10.1007/s11948-015-9649-x.

Costa, Sandra, Hagen Lehmann, Ben Robins, Kerstin Dautenhahn, and Filomena Soares. 2013. " 'Where Is Your Nose?': Developing Body Awareness Skills among Children with Autism Using a Humanoid Robot." In. *IARIA*, 117–22, 2013. http://hdl.handle.net/1822/26038.

Dautenhahn, Kerstin, and Iain Werry. 2004. "Towards Interactive Robots in Autism Therapy: Background, Motivation and Challenges." *Pragmatics & Cognition* 12 (1): 1–35. doi:http://dx.doi.org/10.1075/pc.12.1.03dau.

Duquette, Audrey, François Michaud, and Henri Mercier. 2008. "Exploring the Use of a Mobile Robot as an Imitation Agent with Children with Low-Functioning Autism." *Autonomous Robots* 24 (2): 147–57. doi:10.1007/s10514-007-9056-5.

Goodrich, Michael A., Mark A. Colton, Bonnie Brinton, and Martin Fujiki. 2011. "A Case for Low-Dose Robotics in Autism Therapy." In *Proceedings of the 6th International Conference on Human-Robot Interaction*, 143–44. ACM. http://dl.acm.org/citation.cfm?id=1957702.

Grandin, Temple. 2006. *Thinking in Pictures: And Other Reports from My Life with Autism*. Vintage: Doubleday.

Greczek, Jillian, Edward Kaszubski, Amin Atrash, and Maja Matarić. 2014. "Graded Cueing Feedback in Robot-Mediated Imitation Practice for Children with Autism Spectrum Disorders." In *The 23rd IEEE International Symposium on Robot and Human Interactive Communication*, 561–66. doi:10.1109/ROMAN.2014.6926312.

Grodzinsky, Frances S., Keith W. Miller, and Marty J. Wolf. 2015. "Developing Automated Deceptions and the Impact on Trust." *Philosophy & Technology* 28 (1): 91–105. doi:10.1007/s13347-014-0158-7.

Huskens, Bibi, Rianne Verschuur, Jan Gillesen, Robert Didden, and Emilia Barakova. 2013. "Promoting Question-Asking in School-Aged Children with Autism Spectrum Disorders: Effectiveness of a Robot Intervention Compared to a Human-Trainer Intervention." *Developmental Neurorehabilitation* 16 (5): 345–56. doi:10.3109/17518423.2012.739212.

Nussbaum, Martha C. 2001. *Women and Human Development: The Capabilities Approach*. Vol. 3. Cambridge: Cambridge University Press.

Paine, Robert. 1969. "In Search of Friendship: An Exploratory Analysis in 'Middle-Class' Culture." *Man* 4 (4): 505–24. doi:10.2307/2798192.

Riek, Laurel D, and Don Howard. 2014. "A Code of Ethics for the Human-Robot Interaction Profession." In *10 Miami: We Robot 2014*. http://robots.law.miami.edu/2014/wp-content/uploads/2014/03/a-code-of-ethics-for-the-human-robot-interaction-profession-riek-howard.pdf.

Robins, Ben, and Kerstin Dautenhahn. 2014. "Tactile Interactions with a Humanoid Robot: Novel Play Scenario Implementations with Children with Autism." *International Journal of Social Robotics* 6 (3): 397–415. doi:10.1007/s12369-014-0228-0.

Robins, Ben, Kerstin Dautenhahn, and Janek Dubowski. 2006. "Does Appearance Matter in the Interaction of Children with Autism with a Humanoid Robot?" *Interaction Studies* 7 (3): 479–512. doi:http://dx.doi.org/10.1075/is.7.3.16rob.

Scassellati, Brian, Henny Admoni, and Maja Matarić. 2012. "Robots for Use in Autism Research." *Annual Review of Biomedical Engineering* 14 (1): 275–94. doi:10.1146/annurev-bioeng-071811–150036.

Scott, Ridley. 1982. *Blade Runner*. Suspense/Thriller Sci-Fi/Fantasy. Warner Bros. www.warnerbros.com/blade-runner.

Sen, Amartya. 1989. "Development as Capability Expansion." *Journal of Development Planning*, no. 19: 41–58.

Wainer, Joshua, Kerstin Dautenhahn, Ben Robins, and Farshid Amirabdollahian. 2014. "A Pilot Study with a Novel Setup for Collaborative Play of the Humanoid Robot KASPAR with Children with Autism." *International Journal of Social Robotics* 6 (1): 45–65. doi:10.1007/s12369–013–0195-x.

7 Should You Buy Yourself a "Friend"?

Ethics of Consumer Markets for Robot Companions

1. Introduction

The past two chapters have focused on the relevance of Aristotle's analogy between false friends and false coinage, to two different therapeutic applications. In both cases, the presumed beneficiaries of social robots were likely to be cognitively compromised such that some degree of paternalism seemed to be appropriate; the challenge was figuring out how to promote patients' well-being without being *inappropriately* paternalistic, using a theory of the good of friendship to help guide decisions about what uses would count as actual rather than merely-apparent good.

But this approach has some obvious limitations. In particular, it is not equipped to answer ethical questions about the use of companionate social robots by adults presumed to be rational and generally capable of choosing for themselves, especially in the social arena.

Therefore, I now turn to the issue of consumer markets for companionate robots. Before I begin, some caveats are in order.

First, to a large degree these markets are largely speculative at present. Many robots are being manufactured with companionate features, from iRobot's Roomba series, which communicates to its household with a cheerful series of beeps and short songs, to "smart" toys like Cozmo or NAO, to a variety of sex robots, much-discussed in the news. But few are marketed specifically as electronic friends, outside of the therapeutic contexts discussed earlier.

However, with increasing attention devoted to both consumer robotics and emotionally expressive and responsive robots, it seems reasonable to suppose that there will be at least some attempts to explore the possibility. This could arise in at least two ways: either by designers explicitly attempting to build companionate robots for the purpose of serving as friends to consumers, or by consumers coopting robots designed for other uses, in order to enjoy them as "friends". Should either or both of these things occur, I argue here, Aristotle's analogy will become relevant here in a new way.

My aim here is to focus as closely as possible on companionate robots' relevance to *friendship*. This means, of necessity, that I will be focusing on real and anticipated robots' features only to the extent that they involve friendship. Many robots will also, doubtless, raise issues about domestic labor, sexual ethics and sexual politics, early childhood development, labor markets, environmental impacts, and a host of other rich ethical topics that unfortunately fall outside the scope of the current volume. However, it is my hope that by shedding light on this facet of their potential, the tradeoffs and benefits involved in an overall assessment of such robots can be more accurately assessed.

My other restriction may be more counter-intuitive, but: despite the overlap between toys and companionate robots, and the obvious relevance of friendship to children, I will not be focusing on robot friends for children. My reasoning here is twofold. First, by focusing on friendship for adults, we learn something important about what childhood socialization and friendships aim at. Even if this is not the only good, even if there are goods of friendship unique to childhood, it is surely true that at least one important part of childhood friendship is that it lays the groundwork for rich adult friendships, and so by examining that which it is expected to support, we can work out at least some ethical implications for childhood robot friends. And second, childhood introduces a complicated array of developmental issues that are, for almost all children, in near-constant flux. These complicate the ethical landscape for childhood robot friends without necessarily illuminating the *friendship* component of this technology's presence in their lives. And lastly, children are by definition the population for which paternalist considerations are appropriate. The root of *paternalism* is *pater*, or father/parent. Two chapters of this volume have already been devoted to exploring various paternalist-flavored considerations about companionate robots. What is missing, so far, is an analysis of what happens when we switch from trying to figure out what to do to promote *others'* good, with a sometimes-vague view of what they take that to be, to what is involved in figuring out our *own* good with respect to companionate robots, once all the removable obstacles and obscurities have been removed.

2. The Case for Capitalism

Blay Whitby writes that, when it comes to robot companions,

> [t]here are social dangers that ought to be avoided and about which . . . unconstrained commercial interests may need to be made aware. It is unlikely that the social problems of robot ethics will be solved by allowing markets to decide freely.
>
> (Whitby 2011, 234)

But there are at least two kinds of reasons to think that we ought not to restrict a consumer market for robot "friends". One reason is quite general. Drawing on the work of John Stuart Mill, who argued in *On Liberty* that social interventions on individual behavior should be limited, one might think that individuals are best positioned to judge for themselves what is good for them, and that others' ability to judge one's own good will necessarily be impoverished by comparison.

Mill writes that rational adult individuals are best positioned to decide on matters concerning their own well-being for two reasons. First,

> the person most interested in his own well-being, the interest which any other person, except in cases of strong personal attachment, can have in it, is trifling, compared with that which he himself has; the interest which society has in him individually (except as to his conduct to others) is fractional, and altogether indirect.
>
> (Mill 1975, 71)

That is, the individual in question has the strongest and most direct interest in getting decisions involving personal well-being made correctly. It would be silly to put others in charge of a person's well-being on the grounds of strength of interest. But this seems open to counter-examples: perhaps some people simply do not care very much about their own well-being, and others (especially others who care about them) may care at least as much, as even Mill admits. But the second reason is both subtler and harder to refute, being grounded not on strength of attachment but on access to difference-making details:

> with respect to his own feelings and circumstances, the most ordinary man or woman has means of knowledge immeasurably surpassing those that can be possessed by any one else. The interference of society to overrule his judgment and purposes in what only regards himself, must be grounded on general presumptions; which may be altogether wrong, and even if right, are as likely as not to be misapplied to individual cases, by persons no better acquainted with the circumstances of such cases than those are who look at them merely from without.
>
> (Mill 1975, 71)

That is, when looking at market regulations of social robotics, such regulations will necessarily be based upon statistical data about the impact of robots on individuals that admit of exceptions. For example, suppose a study shows that 80 percent of subjects find themsleves to ultimately be unsatisfied by robotic companions. Potential purchasers of robot companions would be strongly motivated to spend their money wisely.

Money that could be spent on a robot could also be spent out on dinners with human friends, travel, clothes, entertainment, or basic necessities. While the rest of society might also want any given person to be happy, this desire would be mediated rather than direct, and grounded on far less information on the details of a particular individual's life.

This difference in both motivation and relevant information is crucial. For if the study shows 80 percent are dissatisfied, this still leaves room for 20 percent to be satisfied. And the individual consumer has more access to the details of life demands, preferences, and tradeoffs than the market regulators, who necessarily must adopt a generic approach to regulation. So ruling out purchasing a robot companion based on such data could well end up harming more people than it helps, for many of those 80 percent might decide against the purchase because they, along with the regulators, correctly judge that it is not in their interest to do so, while the 20 percent who might be satisfied would be much more likely to purchase these robots to begin with. But they would not be able to incorporate the details that make their circumstances atypical into the judgment of the regulators. So, those who could make most use of them would be deprived of the option, while those most likely to be harmed would be protected against an option they were unlikely to choose to begin with. Thus, making information about such studies available but leaving purchasing power in the hands of the consumers will make for fewer mistakes than the alternative options.

Mill's worry, in fact, is deeper than merely that using statistical guidance on social policies admits of exceptions. He points out that our moral judgments are colored by our own experiences and preferences in a way that makes us particularly ill-suited to judge for others. When our interest in another (particularly some generic other toward whom we feel no strong attachment nor know many intimate details) conflicts with our immediate self-interest, we are likely to miscalculate the effects of their action, both by putting a thumb on the scale in weighing our distaste at their activity against the harm they suffer (as when the religious zealot, says Mill, is offended that others practice a different religion). "But there is no parity between the feeling of a person for his own opinion", Mill argues,

> "and the feeling of another who is offended at his holding it; no more than between the desire of a thief to take a purse, and the desire of the right owner to keep it. And a person's taste is as much his own peculiar concern as his opinion or his purse.
>
> (Mill 1975, 78)

More stealthily, we can mismeasure the badness of an action by projecting our preferences onto the decision instead of legitimately leaving room for others. This, it seems to me, is Coeckelbergh's concern about measuring the value of something like play by using pre-established conceptions of what constitutes play as the yardstick, rather than by observing how playfulness is shaped by culture, environment, and technology. The fact that *I* might not find a robot companion enjoyable is not sufficient grounds for concluding that my neighbor will feel likewise.

Together, then, these present a powerful set of reasons to be skeptical of attempts to judge for others, in the form of market regulations, the appropriate scope of robot companionship. But they are highly general reasons, and those opposed to the laissez-faire capitalist and libertarian philosophy exemplified in *On Liberty* are unlikely to find this completely satisfactory, for a variety of reasons. Surveying the debate here would take us too far afield, so for now I will merely assume that some people are unlikely to be moved by such a dismissal of concerns about robot "friends".

However, in addition to these general reasons to be skeptical of attempts to regulate consumer markets on the grounds that individuals may choose badly for themselves, the particular case at hand offers additional, context-specific reason to be cautious.

This is the idea that a person's choice of friends seem highly particular to the individuals in question. As I have argued earlier, friendships are neither free from generic considerations nor immune to considerations of virtuous character. But they are, nonetheless, so highly individualistic that it would be folly for one person to try to manage another's friendships. In fact, there are informal social norms against doing so. If my friend Alice does not particularly enjoy the company of my friend Betty, then, absent some overriding reason to do otherwise, the correct thing for Alice to do is let Betty be my friend without being hers. The failure to recognize this feature of friendship is the subject of a popular blog post, "5 Geek Social Fallacies" (Suileabhain-Wilson 2003). There may be exceptions, here. For example, the advice columnist Captain Awkward counsels someone who writes in that "my friend hits women" to reconsider the friendship (Peepas 2017). But barring such extreme examples of viciousness, the wisest counsel seems to be to stay out of others' friendships.

Given the particularity of friendships and the general presumption that others are less well-equipped to judge competent adults' choice of companions than the adults themselves, then, it may seem advisable to be particularly cautious when it comes to general rulings on the characteristics or suitability of potential robot companions. If friendship in general calls for a hands-off approach by others, then it is hard to see why robot "friends" should be an exception without further argument. That is, if the reasons governing good friendship are as complex as I have argued in Chapter 1, then it would be inadvisable to attempt to regulate them via most conceivable market restrictions.

Mill introduced what has come to be known as the Harm Principle, commonly understood as the principle that interference with an individual's activities are only justified when and because they harm others, not oneself (Brink 2016). It is difficult to see why befriending a robot would pose a threat to others. While particular companionate robots that are intended to harm others might be impermissible on such an account (for example, someone might make a robot "friend" that would exact colorful revenge on one's exes!), there seem to be few grounds for thinking that robot companions are inherently harmful to others. At best, one might argue (as Turkle 2011 does) that robot companions run the risk of making us worse company for others. But this seems to be justifiable grounds for interference only if we think one's choice of companions making a person worse company for others is generally grounds for interference, and this seems implausible. As Mill puts it, this would imply that we have a right against others that they be good company to us, and this seems to be a troubling right to endorse. If it is distasteful to imagine managing others' friendships for them, it seems even worse to insist that one's social interactions make one an agreeable companion to others with whom one does not necessarily wish to engage.

3. Objections to the Laissez-Faire Approach

In fact, the dominant objections to companionate robots appear to be those that challenge the starting supposition: that consumers in a consumer market for robotic companions are rational.

We first encountered this in discussing the Enchantment objection, in Chapter 4. There, a number of authors objected to companionate robots on the grounds that they enchant us, by evoking emotional responses as if they were real people, even when we know better. I responded by pointing out that a number of enriching components of a well-lived life evoke such emotional responses, from tearjerker films to comedic novels to adventure games. The fact that a romantic story evokes in us feelings of deep sympathy and pathos seems no reason at all to think that it is unsuitable for distribution via a consumer market that leaves purchasing choices to individuals. Nor does the fact that reading romance novels can give some individuals skewed expectations of real-life romantic relationships seem to be good grounds for preventing normal adults from purchasing and reading them. So the enchantment objection, while intended to show that rationality is impacted in a way that interferes with presumptions that each individual is entitled to choose for themselves whether to engage with social robotics, does not seem consistent with the principles and judgments at work elsewhere in emotionally engaging artifacts.

I am happy to grant that human judgments are affected by the presence of stimuli that read to us as social. A variety of experiments in behavioral economics, for example, suggest that something as simple as the presence of eyespots in a person's visual field can impact factors such as trust, cooperativeness, and generosity (Haley and Fessler 2005; Bateson,

Nettle, and Roberts 2006). But if this is so, then it is relevant whenever and wherever such stimuli are present. That means that our conclusions about people's ability to engage in rational judgments should be consistent across contexts. It seems implausible to think that the presence of eyespots on a piece of artwork on the wall should lead us to think that a person's judgment is impaired for the purposes of entering into a contract or making a purchase, for example. So the fact that social robots can present humanoid features that measurably impact human judgment looks like a poor reason to question people's ability to exercise *good* judgment thereby, at least in ways significant enough to license restrictions on consumer markets. But more tellingly, these features have the effects that they do because they have often enough historically been associated with the presence of other human beings. It is because they (admittedly unreliably) indicate that another person is present that they have this kind of effect on us. So it is reasonable to suppose that the actual presence of a person can have at least as strong an effect on the average individual as many stylized humanoid features. So unless we are prepared to stipulate that the presence of another person not only changes but impairs one's judgment, we should be careful to avoid saying this sort of thing about social robots.

But the objection can be refined by specifying some particular effect of appearances that seems to have a particularly distressing effect on our judgment.

Blay Whitby argues that, specifically, there is danger in creating robots that appear to *like* us. "If people come to believe that their robot or caring system is *really* in love with them, then they will probably be a good deal more likely to describe themselves as loving it in return", he claims (Whitby 2011, 241). (Despite the language of being "in love", in this piece he is discussing companionate robots generally and not romantic or sex robots in particular.) "For this reason, a convincing simulation of love is just as ethically dangerous as anything approaching the real thing" (2011, 241) That is, the appearance of affectionate features in particular, rather than humanoid features generally, makes us, he thinks, peculiarly susceptible to false beliefs. He concludes,

> Even, perhaps especially, if the simulation is not particularly convincing, over-enthusiastic marketing by those who wish to sell such technology may deliberately set out to foster such beliefs. This is not an area where we can trust the free market.
>
> (2011, 241)

This runs together concerns about truth in advertising with trust in individual choices. But even if ordinary truth-in-advertising regulations were enforced against marketers of consumer robotics, one could imagine that the argument would still apply: we are peculiarly vulnerable to the appearance of affection, such that we cannot be trusted to make good choices about it. But here, as before, I think that we cannot use this as a reason to regulate the market without substantially changing practices in other areas, which may seem less plausible when considering less novel products. Unless the appearance of friendly salespeople and advertising mascots is similarly considered grounds for challenging the rational decision-making capacity of *autonomous adults* (not the ethics of using Joe the Camel to market cigarettes to children), then this line of argument seems ultimately unsuccessful.

Rather than worry about our vulnerability to friendly appearances as a generic matter, we might worry about our capacity to become attached to particular social robots. Matthias Scheutz argues that the unidirectional bonds humans form with social robots can be dangerous (Scheutz 2012). After surveying a variety of empirical cases in which even educated users anthropomorphize social robots, from Roomba owners who clean the floor occasionally to give the robot a break, to the Army colonel who halted a test of a landmine-clearing robot's capabilities on the grounds that it was "inhumane" (Scheutz 2012, 211), Scheutz concludes that this anthropomorphism is dangerous to rationality. Even with today's largely task-oriented robots, cases like the ones above show that users become attached to robots and make decisions regarding them that seem to be motivated by this attachment. "It will become even easier and more natural for humans to establish unidirectional emotional bonds with more sophisticated robots, often without noticing, akin to becoming addicted, where one's realization of one's addiction always comes after the fact" (2012, 214).

This comparison to addiction carries a great deal of argumentative weight. After all, addictive drugs are a product for which arguments against consumer markets are widely accepted, on the grounds that consumers' rationality is impaired by the good in question. Scheutz reaches similar conclusions about social robots:

> with more sophisticated robots that are specifically programmed to exhibit behavior that could be easily misinterpreted as showing social emotions such as sympathy and empathy, it will become increasingly difficult for people to even realize that their social emotional bonds are unidirectional.
>
> (2012, 216)

That is, he anticipates that continued appearances as of emotional response will override explicit knowledge that the robots are merely executing programs, and that the effect will become ingrained with sustained and sophisticated interactions.

Again, while I see the concern, I worry that because robots are expected to have this effect on the basis of their resemblance to human agents, we cannot consistently say that human rationality is overridden in human-robot interactions without also committing ourselves to the conclusion that we are likewise irrational when interacting with human agents, especially emotionally expressive ones with whom we have long-standing histories of interaction. That result seems suspect enough that I am reluctant to endorse it as a principle for regulating consumer markets.

Scheutz goes on to identify what he takes to be relevant differences between human-human interaction and human-robot interaction. The latter, he argues, involve vulnerabilities that are not present in ordinary human-human interaction.

"What is so dangerous about unidirectional emotional bonds", he claims, "is that they create psychological dependencies that could have serious consequences for human societies, because they can be exploited at a large scale" (2012, 216). If people find it distressing to be separated from their robots, for example, they could be persuaded to act otherwise than they would all-things-considered prefer by the robot threatening to leave. This is different than human-human relationships, charges Scheutz, because

> unlike human relationships where, under normal circumstances, social emotional mechanisms such as empathy and guilt would prevent the escalation of such scenarios; there does not have to be anything on the robots' side to stop them from abusing their influence over their owners.
>
> (2012, 216–17)

But this is implausible. People threaten to leave in order to manipulate and exploit each other on a regular basis. So, again, the claim that robots present a special problem requiring market interventions to restrict individuals' exercise of free choice is implausible unless also applied to human relationships.

Whitby also pursues another line of argument. Individuals might, on a Millian account, be justified in purchasing companionate robots "if by doing so they harm no one else" (Whitby 2011, 239–40). But that is not, he argues, a given. "Even if they are happier with their robotic companion, the reduction in human contact may make them less socially able, and therefore, not so effective as a citizen" (2011, 240) This additional consequence may, he concludes, justify intervention even on libertarian grounds. "If the practice becomes widespread, then society as a whole may suffer, and morally may be entitled to take steps to prevent this sort of breakdown" (Whitby 2011, 240). But Mill explicitly deals with this sort of case in *On Liberty* (Mill 1975) in discussing alcohol's impact on individuals' social interactions. While he acknowledges that alcohol can have a damaging effect on people's social abilities, he argues that we must wait for a violation of a specific duty in order to intervene, not, for example, a general tendency to be worse company, more selfish, and less thoughtful. To interfere before this would imply that we have a right that others be good company for us, and that seems excessively intrusive.

This concern about the effect of robot interactions on our capacity to relate well to others is explored in Sherry Turkle's *Alone Together* (Turkle 2011). She offers a more rich and detailed analysis of the sorts of harms Whitby imagines may leave the individual happier but their social connections worse off. Specifically, she notes that one of the appeals of interacting with social robots is that they can be cleaner, less messy, less demanding, and more carefully crafted to indulge our desires and preferences than interactions with other humans. In a particularly compelling pair of case studies, she introduces a group of schoolchildren to a robot designed to look and act like an infant, and then introduces the same robot to a grandmother during a visit with her grandchild. The schoolchildren consider providing their grandparents in nursing homes with robots of this type as companions, but worry that the robot is quieter and better-behaved than they are, so perhaps their grandparents will prefer the robot to their human grandchildren (2011, 75). When Turkle gives the robot to the grandmother in the other scenario, the children's fears are realized: the woman begins devoting more and more attention to the baby while becoming more and more dismissive of her vocal, demanding, and energetic grandchild (2011, 116–17).

This seems much more plausible as a concern. If we want to enjoy human companionship, we have to accept other people as they are, not as we would prefer them to be. Robots, by contrast, can at least in theory be constructed so as to stimulate our social responses more purely and directly. But of course some people (children and adults alike) are more demanding than others, while some are more compliant, or simply better suited to our own quirks and preferences. And some things stimulate social responses without the messiness of human interactions, from novels to movies to games. While people can choose to surround themselves with yes-men, or with fictional characters rather than flesh-and-blood human beings, indulging their current desires rather than stretching their emotional and empathetic capacities, they seem to make a mistake in doing so. But it does not seem to be the sort of mistake that makes an appropriate reason for regulating markets in entertainment such as books, movies, and games, nor does it seem appropriate to legislate that people spend time around more demanding and less cooperative people over those who introduce less friction into social interactions.

In each case, we have found that the problems presented by companionate robots are also found in human relationships. Given that friendships are widely considered to be inapt targets both of legislation and legislation-like rules, as well as interference by others even in informal contexts, it is difficult to use these problems to motivate legislative solutions for companionate robots.

4. Self-Deception

I think that part of the problem here is that, while these concerns seem to have some (in many cases, a great deal of) merit, they are unhelpful in the context of thinking about general rules for regulating markets, and in particular whether individual consumers have the right to decide for themselves if they should purchase a companionate robot. Rights frameworks simply might not be the best tool for this kind of job.

In the context of discussing the relative merits of various normative ethical frameworks, Rosalind Hursthouse noted that rights-based considerations may often not be the right tools for the job when it comes to interpersonal relations. "In exercising a moral right", she argues, "I can do something cruel, or callous, or selfish, light-minded, self-righteous, stupid, inconsiderate, disloyal, dishonest—that is, act viciously" (Hursthouse 1991, 235). Rights frameworks are not designed to make human beings better people; that just isn't their function, even though we do have an interest both in respecting rights and in being good and becoming better. This point may be especially relevant in intimate relationships, where rights frameworks, Hursthouse thinks, have limited utility.

> Love and friendship do not survive their parties' constantly insisting on their rights, nor do people live well when they think that getting what they have a right to is of preeminent importance; they harm others, and they harm themselves.
>
> (1991, 235)

The fact that these technologies might pose threats to our emotional well-being, might tempt us to take the easy road when it comes to social interactions and social needs, and might make us less responsive to others seem to be legitimate reasons to worry—but not to interfere with others' choices, or at least not to override them. Instead, these are reasons for *individuals* to make good choices in the context of a consumer market for robotics—and for us to make use of the tools for ethical reflection about the nature and value of friendship already on offer. These resources do not necessarily offer neat legislative suggestions or design principles. But they do engage with the rich particularity of social needs of individuals, and situate concerns about robot companions within these theoretical resources.

To see how this is relevant, let us revisit Aristotle's comment on false friends once again. He writes,

> We might . . . accuse a friend if he really liked us for utility or pleasure, and pretended to like us for our character . . . if we mistakenly suppose we are loved for our character when our friend is doing nothing to suggest this, we must hold ourselves responsible. But if we

> are deceived by his pretense, we are justified in accusing him—even more justified than in accusing debasers of the currency, to the extent that his evildoing debases something more precious.
>
> (Aristotle 1999, 140/1165b10)

The previous two chapters focused on situations in which users were at risk of being deceived by designers' pretense. Companionate robots in nursing homes and therapeutic robots for children with ASDs may fool vulnerable patients into thinking these artifacts are their friends. But on a consumer market, consumers are doubly protected against such deception. First, basic truth-in-advertising and truth-in-labeling guidelines should protect against consumers being deceived as to the nature of what they purchase. That is, under even a laissez-faire capitalist system, consumers should be aware of the fact that they are purchasing *robotic* companions. Second, even if one were to, somehow, lose track of this fact, the idea that one could *purchase* a friend seems inherently mistaken.

So that leaves us with the first caution—that if one takes another to be a character friend when the other is doing nothing to suggest this, the responsibility falls to the individual who makes the mistake. And while this may seem to be the end of the discussion from the standpoint of the market, it is the beginning of the investigation when it comes to the individual making the choice.

It is not a new problem, or a bizarre science-fiction scenario, to find oneself vulnerable to the temptation to make another out to be a more intimate friend than is supported by the facts. As Aristotle points out, we like to think well of ourselves. And when another person is kind to us, it is tempting to suppose that this is because they see and are responsive to our distinctive and uniquely valuable qualities. But it is important that we learn to be more discriminating in our judgments, and to recall that there are many reasons for others to be friendly to us—not just in order to avoid being taken in by sociable robots, but in order to thrive in a social environment that includes other human beings. We are capable, Aristotle points out, of enjoying a wide range of social connections that are, broadly, friendly. He includes in the loose category of "friend" everything from civic friendship, the comraderie of fellow citizens in a civil society, to utility friendships, based on mutual usefulness to one another and common in business and community interactions, to friendships of pleasure, common in social circles and recreational settings, to friendships of virtue, which he cautions us are quite rare but nevertheless the most valuable form of friendship. The mature human being, in order to enjoy all of the rich goods of human sociality, should be capable of appreciating *and distinguishing amongst* all of these forms of connection, in order to value each for what it is, and to respond appropriately to others.

5. Putting Concerns About Robot Companions in Context

The dangers to rationality posed by companionate robots are real—but not unique. They are deeply familiar to us precisely because navigating the influence of emotional responses and affective bonds is crucial to functioning well in interpersonal relationships. But here, the solution is not to cut oneself off from potential friends, or to forego emotional connections. Rather, it is to learn to harbor appropriate expectations, to establish and enforce boundaries regarding what one is willing to give up even for another one cares about, to beware of the temptations of flattery and surface pleasures when they interfere with our ability to function well in our lives.

This requires a person to cultivate such an ability to distinguish and value appropriately. It is important not only for purchasing entertainment technology, but for surviving and thriving in a social world, and for excelling as a social creature. This does not exonerate those who seek to exploit our social urges, as Aristotle points out. In fact, those who use the appearance of friendship to harm or manipulate people are guilty of something morally equivalent to (or even worse than) counterfeiting currency. To some degree, this may even provide grounds for regulating companionate robots insofar as their friendly appearances are used to manipulate people to, for example, alter their shopping behavior or way of life. Or at least, to regulate this when it is not an intended effect of the robot for the purchasers—as for example if a person were to get a robotic companion to assist as a personal trainer that would help motivate them to work out on a regular basis. But the broader issues about the dangers of robot companionship, the ones rooted in satisfaction of social desires, are ones inherent to sociality and not specifically presented by robots, and are better handled by general ethical frameworks situating them in the context of human relationships.

Here, virtue ethics seems an appropriate tool. The goal ought not to be to extirpate sociality from ourselves, or to wall ourselves off from things that might activate our social responses, even in novel or unpredictable ways. Instead, we should focus on developing them so as to serve us wisely and well. And here, turning to discussions of the ethical dimensions of friendship can help us to better navigate social robotics. In particular, because we are talking about how individuals manage their *own* social responses by interacting (or not) with social robots, sorting out appropriate management by reflecting on what we value about satisfaction of social needs turns out to be important.

For example, many commentators to date have expressed concerns about the potential for robots to replace some or all human interactions, for at least some individuals. But reasons for replacement can differ. One involves the idea that having some needs satisfied is a task equally well-suited for human and robot companions. The other is the idea that robots may in fact be *better* at some specific aspects of sociality, or at least at triggering enjoyable social responses in human users. These can both lead human beings to adopt robot companions in contexts where they might otherwise turn to human company. But they can involve different motivations that can be problematic in very different ways.

The idea that robots can do some sorts of things better than human beings does not, on its own, seem problematic. Robots can be specialized and tailored in ways that humans cannot. If what I am looking for is a specific and local function, it can make sense to turn to a robot to fulfill it. As Albert Borgmann argued, there can be problems with a widespread tendency to break down complex processes into discrete tasks and parcel out labor to specialized systems, insofar as this can disrupt complicated but enriching interconnected phenomena, replacing them with fragmented efficiencies (Borgmann 1984). But this is a reason to be cautious about over-arching patterns of distancing and delegation to technology, not to avoid specialization altogether. If I need a quick pick-me-up and my friends are busy at the moment, it might be better for me to play a silly game with a companionate robot than to expect them to drop what they're doing to interact with me. Of course, this depends on the details . . . some people might need to get better about asking for help, and social bonds between people can be enriched by such interactions. But there are surely cases where it's reasonable to find ways to meet your own emotional needs without imposing on others. For that matter, sometimes feelings of loneliness can be alleviated by turning up the heat or putting on a sweater—the neural circuitry governing social closeness and temperature sensation overlaps, and sometimes what feels like one kind of issue can indicate a different kind of problem (Bargh and Shalev 2012). At the same time, to think that addressing an immediate feeling of loneliness or frustration constitutes satisfaction of need for social connection seems to be a category mistake.

In this way, I am less concerned, at least in principle, with robots that exceed human beings at performing certain kinds of social tasks, than those that people treat as indistinguishable. Because in order for people to weigh and compare what robots can provide and what humans can provide, they need to think that the goods are comparable. And while robots can provide many of the behaviors we associate with caring, they cannot value us, and being valued by someone we value is an important element of the best form of friendship. In fact, it is precisely this distinction with which Aristotle is concerned in the quote so extensively discussed here. And it is this distinction that presents a moral hazard in *both* human-human *and* human-robot relationships.

Caution needs to be exercised here. The sophomoric temptation is to say that there is something wrong with showing caring behavior without the rich affection associated with the highest form of friendship, because doing so can cause confusion, or make one a mere "fair-weather friend". But our social worlds are more complicated and less binary than this kind of thinking suggests. A cashier and customer smiling and greeting each other, or one stranger holding a door open for another carrying a heavy item, are not doing anything deceptive or dangerous to sociality. Small talk and cocktail party conversations can enrich an evening without the conversationalists needing to become best friends. And even a "white lie" or exaggerated compliment can have its place in some social interactions. To miss this would be to close ourselves off from both the small enjoyments of social life and the very practical benefits of activating our own and others' social responses. Instead, we need to make room in our social framework for a variety of motivations for individual caring or affectionate behaviors, while keeping track of broader patterns, including a wide variety of legitimate motivations, for others to interact with us as they do. These motives include both rich concern for particular individuals and relationships, as well as general pleasantries and social lubricants. We err both when we try to force friendship, and when we count all social interactions as merely instrumental means to an end.

The reason these get run together, both in concerns about social robotics and in practice, I suspect, is that we are not always terribly good at acting on the values we would reflectively endorse, as opposed to the superficial temptations the satisfaction of which only sometimes correlates with what we value. While it is true that good friends are good to us, and it generally feels good when people are good to us, this does not mean that hearing a hard truth is always pleasant, nor that the person being nice to you always has your ultimate well-being as their top priority (even among people who are basically decent and bear you no ill will). This is where reflecting on the values of friendship can help guide us. Bad reasons to take oneself to have friends, or to seek out friends, include superficial flattery and servile deference. More complicated cases involve associations based on mere pleasure. These are not bad in themselves, so long as the pleasures are not harmful, but not to be confused with more valuable character friendships. Again, we see that the concerns we encountered are not a unique problem for robot friends, but can be fruitfully contextualized in a broader issue of reasoning well about friendship.

This also gives us a self-regarding reason to get our conceptions and identifications of the character of social interactions right: friendships are part of a good life for human beings.

This is why we should worry about replacement of human friendships with robot companions, with equating friendliness with indulgence of our whims, with enchantment in the form of following emotional impulses that do not align with our considered rational judgments about the value of the response, and so on. At the same time, valuing friendship can also give us reason to use social, companionate robots well. When they can be used to support what we value, and not just what we superficially desire, they should not be ruled out. And this, ultimately, is why I am against overly zealous market restrictions. While I think that companionate robots can be used badly, they can also be used well, and the details that make the difference are highly context-specific, most easily accessed by the individual, and while individuals are quite likely to make mistakes (as likely as they are in human relationships) they also have the most immediate and powerful reason to get such decisions right, making them ultimately more likely to make good decisions for themselves than others.

6. To What Should We Aspire?

What would good use of a companionate robot look like? To the extent that it emphasizes companionable features, it would be those that work with our social inclinations to become better people, especially in the social realm. In "Friends by Design", John Sullins argues that we should embrace the "work of roboticists that resist the pedestrian notion of robots as domestic servants and see them instead as a chance for us to design new friends and companions" (Sullins 2008, 144). Although this language might sound inimical to my project, elsewhere he clarifies that things like emotions should "remain iconic or cartoonish so that they are easily distinguished as synthetic even by unsophisticated users" and "must be used to enhance the social world of their users and not isolate them further" (Sullins 2008, 156). Instead, he notes that unobtrusive or inhuman-looking robotics tend to reinforce already-existing bad habits of dehumanizing those who assist us, while amplified or exaggerated features designed to elicit emotional responses from us can help us to be more aware, appreciative, and grateful in our daily lives.

Companionate robots that help us to practice gratitude, appreciation, and awareness can help us to become better at social interactions by helping us to exercise our social responses, which, like muscles, can get stronger with practice. Here, the therapeutic benefits of robots in geriatric care and ASD therapy seem as though they can be generalized: just as Wada and Shibata found that nursing home residents who interacted with the companionate robot began to interact more frequently and with less antagonism toward other residents, and as exaggerated facial cues can help children with ASD to practice responding appropriately in social situations, we can imagine that friendly robots can help us to exercise our friendly dispositions.

Robots that help people to fulfill immediate social needs as part of self care, when circumstances make support from others difficult to secure, can also serve an important role, particularly when used as stepping stones as part of the process of improving one's circumstances.

Here, again, care must be taken. I am not fond of the line of argument that posits that robots could be appropriate companions for people too allegedly unattractive or awkward to interact well with other human beings. While I think there is no obligation for any given individual to be a friend to some particular other, it seems that a society where someone's prospects for friendship are severely limited by physical appearance or accessibility (as when transportation and building access are not constructed so as to allow people of differing abilities to use the space) is one that has serious work to do. It also seems as though people who find it difficult to make and keep friends because they are rude, inconsiderate, or difficult to interact with have reason to work on these issues, not find uncomplaining robotic companions. (Those who do are making a mistake of the sort that Aristotle warned about, by confusing instrumental and intrinsic valuing.) But for an individual badly situated socially, having a robotic companion to temporarily bolster one's mood can help a person to form valuable relationships in the long run, much as reading books or watching movies or playing games can help a person to practice empathy and hold out hope for a better future. In fact, exercising one's social abilities by engaging with robotic companions may turn out to be a new form of self care on the order of reading a novel or watching an engaging television series. Coeckelbergh's (2009) concern that we not pre-judge the practice of friendliness by applying current standards to future opportunities seems salient here. But at the same time, it would be a mistake to say that appearances are all that matter—because the history of concern for friendship is shot through with concerns about the appearance/reality distinction. This is not a fringe concern but a central distinction for appreciating the phenomenon.

At the same time, some of the concerns from earlier chapters about the dangers of false appearances of friendship may apply, in limited ways, to a consumer market too. Free markets structured for allegedly rational consumers can permit exploitation of the vulnerable. Just as elderly, confused, or naive consumers can be taken in by predatory scams via any number of channels, especially those that exploit human contact and social norms and expectations, unethical developers of companionate robotics may prey on the easily confused by using emotional levers to manipulate their targets, in social robotics as elsewhere. And to the extent that they do so, I think Aristotle's analysis of the wrongness of deceiving people via friendly appearances as analogous to counterfeiting currency is exactly right. This does not mean that legislating against such activities, or enforcing existing legislation against them, will be easy. But this is a problem for markets generally, and not one unique to a market for companionate robots.

7. Wrapping Up, Moving on

Thus, we have seen that producing appearances of friendship without human-to-human connection turns out to involve both advantages and disadvantages. The advantages can be enjoyed with fewer disadvantages when a clear theory of the nature of friendship is used as a guide in questions about value. In the next section, we move to a different kind of social technology, one where the existence of human-to-human connection is uncontroversial but where the quality of these connections is contested.

Bibliography

Aristotle. 1999. *Nicomachean Ethics*. 2nd ed. Translated by Terence Irwin. Indianapolis, IN: Hackett Publishing Company. https://books.google.com/books?id=-Mf5XV8q6CgC&dq.

Bargh, John A., and Idit Shalev. 2012. "The Substitutability of Physical and Social Warmth in Daily Life." *Emotion* 12 (1): 154–62. doi:10.1037/a0023527.

Bateson, Melissa, Daniel Nettle, and Gilbert Roberts. 2006. "Cues of Being Watched Enhance Cooperation in a Real-World Setting." *Biology Letters* 2 (3): 412–14. doi:10.1098/rsbl.2006.0509.

Borgmann, Albert. 1984. *Technology and the Character of Contemporary Life: A Philosophical Inquiry*. 1st ed. Chicago: University of Chicago Press. https://books.google.com/books?id=XcxUPwAACAAJ.

Brink, David. 2016. "Mill's Moral and Political Philosophy." In *The Stanford Encyclopedia of Philosophy*, edited by Edward N. Zalta, Winter. Metaphysics Research Lab, Stanford University. https://plato.stanford.edu/archives/win2016/entries/mill-moral-political/.

Coeckelbergh, Mark. 2009. "Personal Robots, Appearance, and Human Good: A Methodological Reflection on Roboethics." *International Journal of Social Robotics* 1 (3): 217–21. doi:10.1007/s12369-009-0026-2.

Haley, Kevin J., and Daniel M.T. Fessler. 2005. "Nobody's Watching?" *Evolution and Human Behavior* 26 (3): 245–56. doi:10.1016/j.evolhumbehav.2005.01.002.

Hursthouse, Rosalind. 1991. "Virtue Theory and Abortion." *Philosophy & Public Affairs* 20 (3): 223–46. doi:www.jstor.org/stable/2265432.

Mill, John Stuart. 1975. *On Liberty*. Edited by David Spitz. A Norton Critical Edition. New York, NY: W. W. Norton.

Peepas, Jennifer. 2017. "Our Friend Hits Women." *Captain Awkward*. https://captainawkward.com/2017/05/11/960-our-friend-hits-women/.

Scheutz, Matthias. 2012. "The Inherent Dangers of Unidirectional Emotional Bonds between Humans and Social Robots." In *Robot Ethics: The Ethical and Social Implications of Robotics*, edited by Patrick Lin, Keith Abney, and George A. Bekey, 205–22. Cambridge, MA: MIT Press.

Suileabhain-Wilson, Michael. 2003. "*Five Geek Social Fallacies*." www.plausiblydeniable.com/opinion/gsf.html.

Sullins, John P. 2008. "Friends by Design: A Design Philosophy for Personal Robotics Technology." In *Philosophy and Design*, 143–57. The Netherlands: Springer. http://link.springer.com/chapter/10.1007/978-1-4020-6591-0_11.

Turkle, Sherry. 2011. *Alone Together: Why We Expect More from Technology and Less from Each Other*. New York, NY: Basic Books. https://books.google.com/books?id=J2ine5sIIkgC.

Whitby, Blay. 2011. "Do You Want a Robot Lover?" In *Robot Ethics: The Ethical and Social Implications of Robotics*, edited by P. Lin, K. Abney, and G.A. Bekey, 233–49. MIT Press. https://books.google.com/books?id=oBb-lt3l4oYC.

Part III

Social Media

8 Humans Aren't Cows

An Aristotelian Defense of Technologically Mediated Friendship

1. Introduction

In Part II, I examined ethical issues involving technologies that give us appearances that are close enough to face-to-face interactions to trigger many of our ingrained social reactions, without providing the human connection that has historically accompanied these appearances. In Part III, I turn to technologies that provide real human connection, but in forms that are often novel. What should we think of our connections through social media? In this section, I provide a detailed answer to this question. I begin by establishing some grounds for thinking there is reason to be positive about our prospects here, while at the same time clearing away a number of common objections.

Many have objected, over the years, that technologically mediated friendships are not "real" friendships, or at least are inferior to the face-to-face kind. This strikes me as a mistake. First, the data suggests that for most of us, the distinction between face-to-face and technologically mediated friendships is incoherent, since most interact with their friends in both modes. But furthermore, the reasons used to criticize the quality of technologically mediated friendships do not seem to stand up to scrutiny. In this chapter, I defend the potential for excellent friendship via technological mediation against a number of common objections by appealing to an Aristotelian theory of friendship. I begin by reviewing Aristotle's discussion of the nature of excellent friendship, and applying his conception of the shared life to the activities possible through technology. I argue that excellent, fulfilling friendships may be instantiated in such interactions, because friends can share distinctively human activities such as conversation and exchange of thoughts, mutual development of ideas, making art, and playing games. Aristotelians should thus conclude that the best kind of friendship is available via these new media.

I then engage with and respond to six objections: the privacy objection, the superficiality objection, the commercialism objection, the deceptiveness objection, the physicality objection, and the poverty of communication objection. Having addressed these concerns, I explore some ethical issues for technologically mediated friendship, and conclude by suggesting several ways to use the conception of the shared life to evaluate, design, and use social communication technologies so as to facilitate sharing of flourishing lives.

This line of argument is important for two related reasons: first, despite skepticism (such as that found in Cocking and Kennett 2000; McFall 2012; Fröding and Peterson 2012), it shows that friendships conducted predominantly through computers and smartphones may qualify as the best sort of friendship. Second, it shows that the digital component of friendships taking place both in-person and remotely (which, according to empirical research, constitute the vast majority of social media users' relationships) is at no significant disadvantage relative to the face-to-face component; digital interactions are more than mere stopgaps tiding us over between in-person interactions.

There are a variety of ways of keeping in touch via social technologies. In this section, I will focus on social media, as the subject of a variety of cultural and ethical discussions. Social media has drawn a great deal of attention as we attempt to understand technology's impact on modern relationships, and although specific social media platforms vary widely, there is enough distinctive about them for the genre to deserve specific consideration.

People can experience great emotional attachment to friends even when most or all of their interactions are digitally mediated, and this alone might seem to be grounds for rejecting arguments that excellent friendships are difficult or impossible via social media. But appeals to subjective experience leave open the possibility that such individuals are mistaken about the quality of their friendships; it would be preferable to explain why and how mediated friendships can be close and fulfilling, as part of a unified theory of friendship, and help avoid the charge that people who find such interactions emotionally fulfilling are making some sort of mistake.

2. The Shared Digital Life

As we have seen, Aristotle divides friendships into those based on usefulness (utility friendships), pleasure in each other's company (pleasure friendships), and valuing each other as good in themselves (virtue friendships, which are the best form of friendship, but take the most time to develop, and are the most rare—because, as Aristotle points out, good people are relatively rare, compared to people we might find useful or pleasant.

This does not mean that virtue friends are useless or unpleasant. Good friends are good to each other. Furthermore, many virtue friendships begin as lesser friendships. It takes time and repeated interactions for people to become familiar with each other's characters. The lesser forms of friendship make excellent opportunities to become better acquainted with people. What distinguishes the highest form is not whether or not a friend is pleasant or useful, but whether these traits exhaust the value of the relationship.

Throughout this book, I have made the case that close friends share identity by jointly composing a friendship. These so-called virtue friends have special value, on Aristotle's account, because they share their lives with each other, and in doing so, each becomes "another self" to the other (Aristotle 1999, 1166a30, 1170b5, 1171b30–35). By expanding a sense of self beyond one's immediate concerns, a person can enjoy more good life than is realizable alone (Aristotle 1999, NE1170a-b). The shared life is thus crucial to an Aristotelian theory of the best form of friendship, and my organism account of friendship provides a way to unpack what shared living consists in, on an abstract level. But this leaves open the question: what are the limits of such entities? What kinds of interactions and interdependence are necessary, in practice, so as to achieve shared lives?

In one sense, social media and other forms of interaction via digital technologies seem antithetical to shared living, because they do not (or need not) involve physical cohabitation—one can have such interactions with a person across the globe as easily as the person in the next room. But in explaining what he means by the shared life, Aristotle clarifies that for human beings, living together is accomplished by sharing "conversation and thought", not "grazing in the same field, like cattle" (Aristotle 1999, 1170b10–15). Any medium allowing friends to share conversations and thoughts should thus be compatible with virtue friendship.

The best form of friendship is found between people who share both the broadly human good of rationality, by engaging in collective reasoning through conversation, as well as more specific goods valued by the particular friends:

> whatever his existence means to each partner individually or whatever is the purpose that makes his life desirable, he wishes to pursue it together with his friends. That is why some friends drink together or play dice together while others go in for sports together and hunt together, or join in the study of philosophy: whatever each group of people loves most in life, in that activity they spend their days together. For since they wish to live together with their friends, they follow and share in those pursuits which, they think, constitute their life together.
>
> (Aristotle 1999, 1172a1–7)

The shared life must, then, include both sharing of broadly human good and of particular goods for particular human beings. It is in virtue of this rich and personalized sense of a shared life that we ought to understand friends as other selves.

The shared life in the relevant sense cannot plausibly mean living together as physical cohabitation, hence the caution about differentiating "living together" in the friendly sense from "living together" in the cattle sense. Friends, rather, "live together" by sharing rational life, by sharing language and thought through conversation, as the rational life is characteristic of human beings and conversation is how we reason collectively. In addition, different people find meaning and value in their lives by pursuing different types of personal projects. Individuals share lives with their friends by sharing these activities together: by drinking together, gambling together, working out or playing sports together, hunting together, philosophizing together. In such activities, a friend is not merely valued as means to an externally defined end. The workout partner is not valued merely as pleasant company on runs, the fellow philosopher is not valued merely for providing feedback on drafts in progress and helping work out ideas. In fact, if we are to take seriously the idea that shared friends can complement each other's differences and not just reproduce similarities, we need a different kind of explanation. Rather, a person is perceived by the friend—and the person reciprocates, perceiving her in turn—to be sharing in the valued activity and the valuing, thus expanding each friend's capacity for enjoying worthwhile activities.

In evaluating the potential for friendship over social media, then, we have to look at the sorts of conversations and thoughts exchangeable there, as well as the sorts of activities which could plausibly be shared via such technologies. Can philosophically minded friends discuss philosophy using social media? Can photography enthusiasts share and critique each other's photos? Can gamers play together? Recall that, even for activities which are more difficult to share via computer or smartphone, it can still plausibly be important to individuals who value such activities to reason about them, as reasoning is a characteristically human activity, and conversing about other activities allows people to reason about them together. Sharing a conversation about one's day with a friend should count as living together, if we are to take his comments on the nature of the shared life seriously. Friends need not be present for every life event in order to share in a life: they needn't be grazing in the same field, like cattle. Our capacity to share our lives and thoughts via language and other symbolic representation and artistic expression surely plays a role in the kind of sharing of ideas, experiences, and perceptions that constitute the relevant sense of living together, and conversations are facilitated rather than discouraged by many social media. We should thus conclude that virtue friends may use social media to share their lives.

Another popular use of social media is to share pictures, photographs, music, and videos. Owing to technological considerations, many of these artifacts were not, for Aristotle, goods that he could have imagined people sharing, at least without directly sharing space. In ancient Athens, friends would have needed to attend the same play at the same time in order to enjoy it together, or listen to the same musical performance. But with today's recording technology, in conjunction with the capacity to share and exchange recordings, photographs, and so forth, friends can experience the same events from different times and places, as part of their friendship: a friend might send me a link to a song he enjoys and thinks I would like to hear as well, even though we are in different parts of the world. Sharing such a song can be a way for us to expand our mutual capacity for enjoying good music; if I am confident that my friend values music as I do, I will find it intrinsically valuable that she listens to good music. The popularity of social media for exchanging links to movies, music, and art suggests that we find it important to share these experiences with others via nonverbal means not imagined by Aristotle. However, sharing such goods in this way seems consistent with the spirit of his characterization of living together as sharing human activities: sending a friend a link to a song, for example, strikes me as a human, not a bovine way of sharing the experience of listening to music we both find valuable and enjoyable.

Yet another common use of social media is for playing games. Although this is not the primary focus of the current chapter, it deserves a mention because of the popularity of casual games, from Words with Friends to Farmville, among social media users. Online gaming is a major way that people play games together, and is valued, among other things, as a way of maintaining social relationships (Wohn et al., 2011), so such gaming looks like another plausible candidate for the shared life (Munn 2012). Games are the basis of numerous friendships and playing games together is a shared activity which many people think gives their lives meaning. It engages many of our characteristically human capabilities: playfulness, strategic reasoning, creativity, and exploration. Now, of course, many gaming relationships will be instrumental. People might be friendly in such games merely as a means to greater personal success at the game (as a source of lives in Candy Crush, for example), or for the pleasure of playing together without necessarily valuing the other person's gaming as the activity of one's "other self". But it is not what friends do, but how they value what they do, that makes the crucial difference for distinguishing virtue friendship from the lesser forms. Although some friendships between game players will thus be merely instrumental friendships—friendships of pleasure or utility—others can be virtue friendships, provided friends value each other's lives and activities in the right way.

3. Objections Considered and Refuted

While people rarely think that virtue friends must physically cohabitate, many doubt that excellent friendship can be realized via digital interaction. If these criticisms were correct, they would cast doubt on my claim that friends can plausibly "live together" via technological mediation in its most popular social forms.

If friends cannot trust each other in their digital communications, or cannot share their valued activities owing to a lack of privacy, cannot interact with each other as particular individuals rather than through mere superficial broadcasting of interests and life events, or if their relationship is adversely affected by the commercialism of social media, or if the best friendship involves a kind of physicality impossible to experience in digitally mediated interaction, or if the communication available via social technologies is impoverished so that conversations that take place by such means do not allow friends to enjoy sufficiently rich exchange of ideas to allow friendship to flourish, then my argument will have been insufficient to show that social media supports excellent friendship.

I therefore consider and refute six objections to my thesis: the privacy objection, the superficiality objection, the commercialism objection, the deceptiveness objection, the physicality objection, and the poverty of communication objection.

3.1. Privacy Objection

Because social media often facilitates highly public conversations, the intimacy characteristic of the best friendships would seem impossible on most common social media sites. Friendships cannot thrive without the capacity to confide secrets (Thomas 1987). Otherwise-friendly exchanges on Twitter, for example, seem much too public to allow for such intimacy. Even if one's privacy settings are relatively locked down, the exchange would still be visible to all of one's other contacts.

My response to this objection is two-pronged. First, many social media include one-on-one communication channels as well as broadcasting platforms. So most social media developers already recognize the importance of private communication channels. This may not be much consolation, however, on its own, as one might think this still leaves some the most distinctive parts of social media—many of those features that distinguish them as communication channels from older technologies such as email or SMS messaging—inimical to friendship.

However, the second portion of my response is to point out that not every interaction between friends plausibly involves secrets. Conversations conducted publicly may fail to allow for the exchange of secrets, and even convey an implicit invitation for other interested parties to join in, and simultaneously constitute important parts of a friendship. Friends can socialize at parties, making their conversations implicitly open to other party-goers, and such conversations may constitute rich and fulfilling sharing of lives between friends.

Secrecy may be necessary for friendship in the long run, but it does not follow that every exchange in a friendship must be conducted in secrecy, or that the public parts of a friendship are any less important to the friends or the well-being of the friendship. Unless we rule out social and semi-public settings more generally as appropriate grounds for friends to interact, we cannot consistently criticize social media on this basis, and the more general criticism seems highly implausible.

That said, social media services seem increasingly sensitive to user desires to communicate with smaller groups of friends or one-on-one via enriched means. From disappearing direct messages, to a wide array of stickers, easy sharing of video clips and animated gifs, to doodles and photos, to the ability to create private pages and groupings for small groups of people to discuss and share, private and small-group messaging capabilities are increasingly featured in platforms from Facebook to Instagram to Snapchat. Monitoring user adoption of these channels may offer us valuable information about the extent to which people's relationships thrive in public versus private spaces (and the wide range of contexts that do not fit neatly in either extreme).

3.2. *Superficiality Objection*

According to the superficiality objection, on social media one simultaneously addresses everyone one knows, and no one in particular, and so such exchanges are superficial; rich conversation with particular friends, of the sort which would constitute sharing a life, is difficult or unlikely (Bloomfield 2015). More directed address of particular friends would call on and invite their unique perspectives and contributions, but a broadcast to all one's acquaintances on some social media network is necessarily so thin and watered-down that the richness of personalized exchange is lost.

My response is that this presumes one (perhaps common) way of using social media as the option, or at least the default option. As noted above, a growing array of alternatives connects individuals and small groups without exposing them to the public eye. But if something is wrong with the broadcasting approach, and I am sympathetic to concerns to this effect, this at best constitutes a criticism of people's use of the medium, not the medium itself.

A better model for interaction on social media seems to be something like the following: public postings may be directed at friends with a shared interest, even while conveying an implicit invitation for others to join in the conversation. These directed addresses may be quite specifically tailored to call on a friend's shared interests, character, sense of humor, and so forth, and simultaneously convey an implicit invitation for the like-minded to join in. Although public, links, stories, pictures, and videos can be shared because one believes a particular friend would like them, find them interesting, want to respond to them, and so forth, and the friend could plausibly accept these friendly tokens without finding them diminished for being communicated in a public forum.

Suppose, for example, that I post pictures of a cooking endeavor in progress to Twitter, so that a couple of my friends who share my interest in cooking can contribute suggestions, answer questions ("Help! Why is this dough taking so long to rise?"), and share in the experience. (Some of these friendships may be instrumental, but if cooking is valued by the friends in the right way, sharing my cooking experience with you may be part of our shared life and hence part of a virtue friendship.)

Because this conversation is public, another acquaintance may thereby realize we share a hobby and join the discussion. But it is not obvious that this cheapens or detracts from conversation with my established fellow cooking enthusiasts, any more than would an equivalent conversation at a public gathering where the acquaintance joins in.

Thus, even broadcast social media poses no intrinsic threat to the richness and particularity of friendly interactions. There may be some temptation toward making superficial broadcasts, but there are alternatives available. At best, we end up with a normative claim about how users ought to interact via social media, rather than a condemnation of the medium. Users will be best positioned to enjoy excellent friendships if they show appropriate sensitivity to social context and some skill at handling complex social exchanges, but this is not so different from the skill set which benefits any other friendship.

3.3. Commercialization Objection

A number of prominent social media platforms are commercial in nature. Companies such as Facebook and Twitter make money off of their users in a variety of ways. Most prominently, purveyors of free-to-users networks tend to make their money through advertising: the users are, at the end of the day, the product (a dedicated audience, about whom the company knows quite a bit of information which advertisers find valuable as a means to more effectively target their market), being sold to advertisers who are the ultimate consumers of such networks. Users of commercial social media, arguably, ought to know and take into consideration this arrangement, but one still might wonder whether friendships conducted in such environments are ethically problematic—are friendships somehow weakened, harmed, or made less valuable if they are the means by which the companies providing the space earn profits?

In order to effectively address this challenge, it is helpful to distinguish several potential problems commercial social media might pose for friendship. It might be the case that friendships conducted under such circumstances, for example, are qualitatively inferior to those in which the friends do not serve as sources of revenue in virtue of their friendship—in something like the way that "diet" versions of food are often qualitatively inferior to those made with plenty of butter and sugar. It might be that friendships on social media are at no intrinsic disadvantage, but that companies which profit from such friendships tend to harm the friendships in other ways, perhaps by snooping, violating privacy, and interfering with communication channels, as when Facebook changes the items displayed in its News Feed (a non-exhaustive display of updates from contacts on one's "Friends" list). It could be the case that friends harm each other by encouraging their friends to serve as products for such companies to sell. Or, upon closer examination, the problem of commercialism might not look so terrible after all.

Tackling these in order: consider the first challenge: that friendships on commercial social media are weaker or less valuable than their equivalents elsewhere. It is implausible to me that friendships and friends' interactions are intrinsically qualitatively less when and because other people profit from them. If I go to the bar with my friends on Trivia Night, the bartender and bar owner benefit from my friendship, but it is implausible that this weakens or lessens the friendship.

Someone might object that this is disanalogous, because friends at a bar are not the product; they are the consumers. But other scenarios can be constructed in which friends engage in activities from which others profit, even ones in which the friends are products rather than consumers, and where friendships do not thereby seem weakened. Suppose my friend and I together perform at a local coffee shop's open mic night. The coffee shop owner holds such events because they attract paying customers, even if the performers purchase nothing: they are the product being used to sell the coffee shop and its related goods to other consumers. And yet writing songs and rehearsing them, with the goal of eventually performing them at an open mic, does not seem intrinsically less friendly than the aim of writing and performing them for some other end.

Or, consider the practice of holding Ladies Night at bars: in this business model, the establishment offers free drinks to women, in the hopes of attracting more male clientele, said men being attracted by the opportunity to make the acquaintance of the women attracted by the free drinks. If a woman and her friends decide to take advantage of Ladies Night as an opportunity to socialize, they do not seem to be at any particular disadvantage, at least so far as their friendship goes. (They may or may not advance their other interests, depending upon their tastes, preferences, and desires, as well as the conduct of the other patrons of the bar.) In this case, the friendship does not seem to be threatened or lessened in virtue of the bar owner profiting from their use of the goods offered in hopes of attracting women to the bar, in order to sell goods to someone else. So I conclude that it is implausible that friendships on commercial social media platforms are intrinsically lesser than their nonprofit counterparts.

I take much more seriously the charge that friendships can be harmed by violations of privacy, and the trust friends place in each other. It is highly plausible to me that one of the bad things about, for example, government surveillance is the damage it does to friendship. But there is an important disanalogy between state surveillance and the data monitoring in which commercial social media companies engage; the voluntary nature of friends' involvement. If the government eavesdrops on you, they do so without your consent, and possibly without your knowledge. If friends are aware that their conversations and exchanges may be monitored, have other options (such as email and noncommercial social media networks) for interacting, and yet choose to engage with each other in this setting, fully conscious of the circumstances, that seems to me no bigger a threat to their friendship than having a conversation in a public space where others can eavesdrop—such as at a bar, coffee shop, restaurant, or public park. If social media providers engage in deception about the type or scope of their information gathering and monitoring practices, so friends do not make their choices in full knowledge of their lack of privacy, then I think the company does wrong—and there seems to be evidence that companies do engage in such deception on occasion. But I do not think this need be the case in commercial social media, and it is important to distinguish specific harms of wrongful action from generalized threats intrinsic to the medium.

The last possibility also seems to me to be defused by emphasizing the voluntariness of participation in commercial social media. Friends may choose to interact via social media, but friends who try to force others to do so do wrong—not because of something about social media, but for failing to respect their friends' agency, for attempting to coerce them, and for failing to take seriously their friends' reasons for preferring not to engage. Note that the last is distinct from concerns about whether these friends are correct in preferring not to use any particular form of social media—as I have argued in Chapter 3, good friendship seems to require some willingness to treat the friend's subjective interests as valuable, simply in virtue of their being the friend's, and so friends who fail to treat their friends' subjective interests as valuable are thereby failing at being good friends.

Because so many of my responses appeal to voluntariness, it is entirely possible that those traditionally considered incapable of giving voluntary consent, particularly children and youths, may not be well-suited to make good decisions about their social media usage, and this may have impacts on what children should be permitted to agree to—or what companies can do with their underage user base. But at minimum, adult friendships conducted on commercial social media emerge relatively unscathed.

3.4. Deceptiveness Objection

It is easy to make deceptive claims about oneself when interacting via a variety of social technologies, either deliberately or by subconsciously editing what one reveals about oneself (Walther 2007; Hancock 2009). Therefore, runs the deceptiveness objection, trust in those friends with whom one's interactions are primarily digitally mediated should be minimal. Or at least, if one trusts one's friends, such trust is ill founded.

A friendship grounded in ill-founded trust seems to be a weaker or lesser friendship than one where trust is merited, one might suppose, and so technologically mediated friendships are bound to be weaker or lesser than their in-person counterparts.

But we must be careful to consider the alternatives: in this case, the problem of deception in face-to-face interactions, and the potential for such friendships to thrive despite this risk. Deception and editing of the self one presents to acquaintances is quite possible—and commonplace!—in person, and face-to-face interactions also present vulnerabilities. Skilled performers and con artists, for instance, exploit people's confidence in their ability to gauge character by expression and tone of voice, so the "mark" fails to pay adequate attention to the content of the con artist's claims. Even in more ordinary and less sinister exchanges, we frequently use voice, tone, expression, and posture to elicit desired responses from our conversation partners and to subtly direct the conversation toward topics we enjoy and away from uncomfortable subjects. I return to the topic of manipulation in face-to-face interactions, and the advantages of technologically mediated communication relative to this, in the next chapter.

Many technologically mediated conversations are, arguably, less susceptible to deception of this kind. Although (as I discuss in Chapter 10) emoji, photographs, and other non-verbal features of computer-mediated conversations can do some of the work of facial expressions and tone of voice, there is less pressure from the immediate presence of the speaker. One can evaluate the content and consistency of claims with a more critical eye. Many mediated conversations leave digital "paper trails", making it easier to cross-check stories and consider a person's comments in light of the overall picture of their character presented by their digital footprint. For example, the person who expresses one view on social issues to you, but whose Facebook wall is full of posts and memes to the contrary, gives grounds for an overall assessment of character which takes the totality of evidence into consideration. Of course, we should take account of the ways that different audiences and social contexts can help frame issues. For example, danah boyd relays a compelling story of a young man whose social media presence was littered with gang signs, but whose college essay expressed a deep desire to get out of his crime-ridden neighborhood. She counseled the university admissions committee to consider the value of publicly signaling belonging to members of his current community, even if he hoped to leave it behind, when weighing the likelihood that he was lying in his admissions essay versus his social media presence (boyd 2014). People are complicated, both face-to-face and otherwise.

It is by no means clear that the grounds for trust in person versus in mediated interactions weigh heavily in favor of one versus the other, and there is good reason to think one ought to exercise good judgment and discrimination in both circumstances. Even if deception is easier or more common via technologically mediated interaction, at best this is a difference of degree, not kind. There is also evidence that people tend to detect quite a bit about even text-based correspondents' intentions and truthfulness via things like word choice and phrasing, which will tend to help mitigate concerns about the ease with which ordinary people can deceive each other through digital channels (Wallace 1999). If the potential for deception in face-to-face interaction is not sufficient to rule out the possibility of friendship, neither should it be considered especially hazardous to technologically mediated friendship.

3.5. Physicality in Friendship

One of the most obvious differences between friendships on and off social media is the possibility—or lack thereof—for physical contact, and more generally, the physical presence of the other person. As social creatures, many of our social needs may be met through digitally mediated interaction—but some may not be. In many contemporary Western cultures, particularly between men, physical touch is not typically recognized as an important component of friendship and is often considered taboo (Strikwerda and May 1992; Derlega et al., 1989; Monsour 1992; Bank and Hansford 2000), but for many people touch can be an important and valuable part of sociality. Furthermore, just being around other people can itself be important to many of us; think about the attraction, for some people, of heading out to a coffee shop to read rather than staying at home, or working on a project at the library rather than in a solitary office.

Clearly, these social goods cannot be realized in technologically mediated friendships, while they are often part of face-to-face friendships, although with the caveat that this will vary depending on the individual and the culture. In Arab cultures, for example, male friends often hold hands, while such behavior is considered highly unorthodox in the US: see, for example, the furor over a US president holding hands with the king of Saudi Arabia (Hogluin 2005). This raises questions about whether members of less physically affectionate cultures thereby miss out on something valuable, and it seems plausible to me that the answer could be "yes".

Does this mean that social media friendships are intrinsically at a disadvantage? Perhaps, but I am not sure that this disadvantage need be severe. In sexual relationships, many (although not all) people find monogamy important, but there is no parallel norm for friendship; in fact, it is often thought to be healthy for people to have different friends who meet different needs, share different interests, and serve as sources of very different goods; in fact, variety in friendship may plausibly be one component of an excellent life. Given this, it is not an easy or obvious transition from the thought that one social good is unavailable in a particular friendship, to the conclusion that the friendship is thereby deficient, nor that one is worse off for having that friendship. So long as one's need for physical contact and sharing space is met (and such needs may vary widely depending upon the individual in question) some particular friend need not be a source of those goods. What may be the case is that people are worse off if none of their friendships involve physical presence or physical contact, but unless technologically mediated friendships replace other friendships in a person's life, they do not thereby make people worse off, and the data so far suggests that mediated friendships do not replace face-to-face ones (Wellman et al. 2001; Ellison, Steinfield, and Lampe 2007; Rainie et al., 2006; Ito et al., 2009). Lastly, some technologies, such as the Apple smartwatch, allow users to engage in physical interaction by, for example, sending each other recordings of their heartbeats that cause the other's watch to pulse in time with the recording. And telepresence robots are an intriguing if still developing way to incorporate the physical into the realm of technological mediation. Time will tell if such innovations will help meet people's needs to share physical interactions, or whether they will remain novelties, while in-person interactions continue to be considered preferable. My prediction is that they will supplement but not supplant in-person interactions, much like voice, text, and video communications, and so will not so much transform my argument as illustrate it in another context. But I am open to being disproved.

3.6. Poverty of Communication Objection

The final objection I consider is what I will call the poverty-of-communication objection. This has changed over time as technologically mediated communication has evolved from simple text-based exchanges to a much richer range of options, but concerns about the quality of communication channels remains. I grapple with this issue in greater detail in subsequent chapters, exploring the various advantages and disadvantages of different communication channels and the ethical issues involving appropriate choice of communication channel. Here, I start with some of the most familiar sorts of objection, to clear the way for more detailed discussion later.

Cocking and Matthews, Vallor, and McFall introduce related objections to the potential for the best form of friendship via digitally mediated interaction (Cocking and Matthews 2000; Vallor 2012; McFall 2012). Although Cocking and Matthews' original concerns were addressed by Adam Briggle, who argued that distance and deliberateness could enhance friendship by facilitating introspection and mutual disclosure (Briggle 2008), later authors have not found this response entirely satisfactory (Vallor 2012; McFall 2012). I supplement Briggle's account by introducing a distinction between sharing information about activities, and sharing the activity of discussing or reasoning about that same information. Making this distinction helps address concerns about the feasibility of sharing lives on social media.

Cocking and Matthews, McFarland, and to some degree Vallor, all criticize the nature of technologically mediated communication. They express concern that something about the nature of this communication makes it impossible, or at best, significantly more difficult or unlikely, for friends to share thoughts and feelings in ways constituting an Aristotelian shared life. This is related to both the deceptiveness and superficiality objections addressed earlier, but is raised often enough to deserve a dedicated response.

Friends may converse about their lives, they charge (it would be difficult to deny that friends do, at this point in time), but such conversation fails to count as the rich sort of sharing Aristotle had in mind. Sharing a picture of lunch on Instagram seems qualitatively different than sharing a meal together in person.

This sort of objection can involve either of two claims: 1) I cannot actually share activities with my friends, only talk about (or otherwise report on) my activities with my friends; 2) Even if such talk constitutes a shared human activity sufficient for Aristotle's sense of the shared life among friends, it is in principle impoverished when conducted via the various communication technologies as opposed to a face-to-face conversation and so is an inferior part of a friendship. For example, in relating something to a friend face-to-face, my expression might give me away at a key point in describing an emotional encounter, but textual communication strips away such supplementary information channels. In the latter case, I merely report my conscious experience and interpretation of my life to someone who can then contribute little to my understanding of the portions of my life for which my friend is not present and of which I was not initially aware (Cocking and Matthews 2000). This seems significant even if one uses supplementary means such as emoji and photographs to provide emotional expressiveness and contextual information lacking, or difficult to supply, in a pure-text report.

I think that neither claim is warranted. Consider the first one first. Vallor expresses it as follows:

> the possibilities for sharing lives online look relatively impoverished if we grasp the distinction between sharing lives and sharing about lives; the former involves performing together the activities that make up a life, the latter involves communicating to one another information concerning our lives, without implying shared activity.
> (Vallor 2012, 196)

But this does not hold up under scrutiny.

Multiplayer gaming is perhaps the most obvious example of a shared activity that is plausibly conducted across a computer network and which could constitute the sort of shared life characteristic of virtue friendship (Munn 2012). Other examples, however, come readily to hand. Journaling has a long-standing foothold in social media, distinguished from private diary-keeping by the fact that such journals are kept with the intention of introspecting on one's life in the company of others (Søraker 2012; Briggle 2008). In addition to older social media platforms such as Livejournal, more recent innovations are often used as a form of journaling, from tweeting thoughts and events as they happen, to Instagramming moments throughout the day, to building stories in Snapchat from short video clips and text-enriched photos.

Events described in a journal entry or its digital equivalents are not directly shared with the friends, any more than the experiences relayed during a conversation transport one's conversational partner to the scene described. However, the activity shared by participants in journaling is not the event related in the journal entry, but the journaling. Journaling is a respectable activity in its own right, valuable for introspecting and reflecting on oneself. It is useful for gaining self-knowledge as well as better control over the direction of one's future.

Writing, photographing, filming, reading about, and discussing one's life is something that can be shared with friends and to which friends can be unique contributors, offering valuable insight on something one might not have noted on one's own. Having a co-perceiver of one's perceptions can greatly enhance one's understanding and development of oneself. Rather than think public journaling consists merely of relaying one's day as one experiences it, to which friends and observers could contribute nothing new, as charged in the objection, journaling is plausibly valuable precisely because narrating one's experiences in order to reflect on them can offer people new insights into themselves and each other.

This use of social media is directly related to character development of the kind Aristotle specifically calls out as characteristic of the best friendships: both appreciation of extant character and the role played by friends in modeling, critiquing, and offering feedback which shapes the development of a friend's character.

Not all friends engage in shared journaling, of course, and other activities may be harder to share via communication technologies; friends who hike together, for example, may find shared journaling about hikes to be a poor substitute for actually climbing a mountain together. But that doesn't show that digitally shared activities are inferior for friends who find these activities—such as gaming and journaling—to be intrinsically rewarding. For comparison, consider again the friends who hike; the fact that they find shared hikes to be an important component of their friendship does not mean that all friends must hike. Conversely, we can easily imagine that another group of friends (perhaps a group of gamers) might find hiking to be unpleasant and, at best, instrumentally valuable for maintaining cardiovascular health and only made bearable by the fact that on a shared hike, the friends can discuss gaming, even though they consider such discussions a poor substitute for actually gaming together.

Conversations, moreover, may be directly shared both face-to-face and when mediated by technology. Conversationalists do not directly share their described experiences with conversational partners, but the conversation itself is a direct activity in which both friends engage, and to the extent that conversing is collective reasoning and reflection on one's life, friends thereby share lives. I suspect this is Aristotle's point to begin with, in distinguishing human "living together" from cattle "living together". In that case, digital communication is just another way (and, in addition, a distinctively human way) to have such conversations.

This brings me back to the second aspect of the poverty-of-communication objection. Many non-verbal cues are present for face-to-face exchanges but missing or modified in textual communication. In Chapter 10, I explore the role of non-verbals such as emoji in technologically mediated communication. But here, I tackle the question: are they inherently inferior to face-to-face expressions and other non-verbal cues? These non-verbal cues are lauded by McFall and Cocking and Matthews for being "involuntary" and "unconscious" and thus (presumably) superior indicators of one's true self (McFall 2012; Cocking and Matthews 2000). But as noted in the previous section, we often exaggerate or consciously control facial expression or tone of voice for dramatic effect. On the other hand, word choice and phrasing can reveal quite a bit about a person without their consciously intending to do so (Søraker 2012; Wallace 1999). Furthermore, Briggle makes a persuasive case that computer-mediated communication, relative to face-to-face discussion, enhances important qualities such as candidness and deliberateness (Briggle 2008).

Thus, conversing still counts as a directly shared activity, and one in which the sensitive and skilled writer and reader can participate fruitfully. Of course, not every social media user is sensitive and skilled, but neither is every conversationalist—one can be tone deaf, oblivious, and so forth, and thereby have one's potential for friendship limited by such handicaps. Meanwhile, generations of prose testify to the potential for rich communication by text. Skills of writing and reading then turn out to be important for developing and maintaining rich social relationships on social media, just as the ability to be both a good speaker and a good listener are important to developing and maintaining good face-to-face friendships. They help us to excel at the activities that constitute friends' shared life, especially when supplemented by graphic and other non-verbal features.

This objection has one last form. My responses all point to speculative possibilities about how people might use social media to connect, one might say. But the data is in: people who use social media frequently appear to experience depression and anxiety (Lin et al., 2016). This might seem to show that these forms of interaction are inferior to face-to-face ones, which tend to boost mood and psychological well-being (Cohen and Syme 1985). But the data actually shows something much more subtle and interesting. Social media users divide into two groups: those who use them primarily passively, to observe others, and those who use it actively, to post, share, comment, and like. The passive users appear to be prone to depression and anxiety, an effect theorized to derive from the comparative nature of the activity (Verduyn et al., 2015; Burke, Kraut, and Marlow 2011). When scrolling through a series of attractive vacation photos, witty quips, and evidence of delicious meals, it is easy to compare one's everyday life to others' highlights and come up wanting. Sustained comparisons with others can be both symptom and exacerbator of depressive and anxious symptoms. But those who use it to connect with others, through active engagement, do not experience this effect, especially when they do so in moderation. Instead, they show evidence of realizing the goods of social connection. That is, when taken as an *interactive* medium, social media options have the potential to reinforce social connections, and this is an especially strong effect for those who use it as one tool among many to reinforce cross-media social bonds. But for those who use it to keep others in view but at arms' length, it is more akin to people-watching and less good for us.

4. Aristotelian Guidelines for Digital Friendship

It is not, of course, my intention to argue that all technologically mediated friendships are good ones, or even that all close mediated friendships are good. In what follows, I explore some ways that friends' actions can badly impact friendships with a digital component.

4.1. When Bad People Team Up Through Technology

Aristotle's account of virtue friendship makes virtuous character central to the best friendship, as we saw in Chapter 3. He holds that people who are close but not virtuous can be actively bad for each other, especially by sharing vicious activities with each other: "the friendship of base people turns out to be vicious", Aristotle argues, "For they [the friends] are unstable, and share base pursuits; and by becoming similar to each other, they grow vicious" (Aristotle 1999, 153, 1172a10). As I have argued in Chapter 3, friends with bad character can be bad for each other in a variety of ways, by sharing bad activities (that is, living badly) together, by influencing each other for the worse, and by reinforcing existing bad tendencies and thus making it more difficult or less likely for their friends to improve. Here I focus on ways that this is specifically important in friendships conducted via social media.

As we have seen, Aristotle thinks the shared life is central to friendship: the quality of the life shared determines the quality of both the friends and the friendship. One's quality of life is at least heavily influenced by one's character, particularly the activities in which one engages and which substantially constitute one's life. The good life (*eudaimonia*) is the life of the virtuous person who has the external goods necessary for such a life, and friends who enjoy the best sort of friendship will share the best sort of life. Given the variety of artistic and intellectual pursuits which can be pursued through communication technologies, as well as pleasant components of a good life such as games and entertaining and enjoyable conversations, it is plausible to think that many excellent activities can be shared through technology and so help constitute excellent friendships.

But of course, not all such activities are excellent, and some are characteristic pursuits of the vicious rather than the virtuous; friends who share in such activities are close but not good for each other. Even though they share lives, the life they share is not worthwhile:

> What is good by nature is also good for the decent person; that is why life would seem to be pleasant for everyone. But we must not consider a life that is vicious and corrupted, or filled with pains; for such a life lacks definite order, just as its proper features do.
>
> (Aristotle 1999, 149, 1170a20)

Taking this in conjunction with the idea that base people share bad activities which would not be part of a good life, we end up with the idea that friends must share good lives in order to be the best of friends. It follows quite naturally that friends who engage in vicious activities across digital networks, whether bullying, plotting terrorist activities, or the unethical forms of hacking, do not live the good life together and so are not good friends, even though they may be very close.

More subtly, but just as importantly, many of the common elements of technologically mediated friendship, such as journaling, conversations, and discussions of various forms, have friends both revealing their own characters and influencing each other, both subtly and directly: from explicit advice and prescriptions through modeling and emulating each other, to reinforcing each other's vices or virtues in discussions and shared deliberations. Friends may thus have great influence on each other's characters in both face-to-face and digital interactions, and if character is important to one's ability to live the good life, then good friends will influence each other in good ways, while bad people will do the opposite and thus be bad friends, whether or not they intend to do so.

4.2. Pressuring Friends to Join or Use Social Media

Friends sometimes feel pressured to join or use social media to keep up with their friends. They may even find themselves left out of invitations and event planning conducted via social media to which they do not attend. Although such social dynamics are no doubt real and problematic, this does not seem to pose any special challenge to social media in particular; friends who share the same space and communication channels are often at an advantage relative to friends with whom one has to go out of one's way to connect.

Good friendships seem as though they can be sustained despite such obstacles, and while I argue here that good friendships are possible via social media, this is not meant to condone attempts to restrict friendly activities to such media when it is inconvenient or otherwise unattractive for one's existing friends to do so. But the problems with the sort of behavior described here do not seem to be the result of social media per se, but rather implicate broader issues about friendliness, cliquishness, inclusivity, and other social challenges with which the would-be friend must wrestle, regardless of technology. It seems intuitively unfriendly to insist, for example, that one's friend accompany a person to an expensive restaurant when the friend is short on cash or trying to save money, but this does not show that there is something wrong with expensive restaurants; rather, it demonstrates that good friends ought to be sensitive to their friends' limitations and other interests, and work to find a common ground.

That said, if social media is an easy and convenient way to keep in touch, it is plausible that in some circumstances the friend who refuses to do so is in the wrong, much as the friend who refused to use phone calls to keep in touch and insisted on making all interactions take place "in person" might, under some circumstances (and here the details would definitely make a difference) be "difficult" in an unfriendly way and place unreasonable demands on the friendship.

4.3. "Blocking" Others on Social Media

Relatedly, some relationships on social media become strained when people confuse standards for legal regulation of communication with issues of personal communication in friendship. Social media programs often give users the ability to block material originating from users whose content they wish to avoid. Although such behavior seems unlikely in characteristically close and high-quality friendship, it can be a helpful way to manage one's contact with looser social connections whose tastes, interests, and politics clash with one's own. There is an interesting side question here about the extent to which the resulting silo effect is bad for us, and the degree to which the virtuous individual ought to remain open to differing viewpoints of less-intimate social connections. But where the line should be drawn will not make a difference for the following argument, and is beyond the scope of this project.

Relational friction can transpire when someone discovers that a connection has blocked them on some or all social media channels. To make sense of these cases, it is important to differentiate kinds of concerns. From a political perspective, when one person blocks another from their news feed, no harm is inflicted on either party: freedom of speech is standardly interpreted as not entailing that others must listen to you, and so exercising discretion in what one chooses to consume on social media does not, plausibly, infringe on others' rights or violate any legal duties or obligations. "Friends" who object to being blocked by claiming something like a violation of their freedom of speech thus miss the point of the principle to which they appeal. However, in close interpersonal relationships, such practices can be used to minimize conflict, sometimes at the cost of honesty and openness. It can be cowardly to avoid difficult conversations with a person, simply by refusing to look at the things they say or do with which you disagree strongly.

However, neither are the blocked parties necessarily blameless; failure to take one's audience into consideration when (for example) endorsing something that belittles or demeans people with whom one wishes to maintain a close relationship seems ill-advised and unfriendly. Deciding whether a comment calls for tactful avoidance or honest and open airing of grievances can be highly difficult and calls for skill and sensitivity to a variety of concerns, including how close one is or desires to be to the other person. This, in turn, can itself be challenging; the boundaries between close friend and loose acquaintance are often blurry in real life, even as we acknowledge differing norms for each kind of relationship. It remains the case, however, that good friends can and should be able to overcome such difficulties without taxing an Aristotelian theory of friendship, or the resources of social media. Instead, characteristics such as tact, forthrightness, and sensitivity to context, courage to be able say the difficult thing, and wisdom to know which fights are worth having and which are not, remain as relevant today as before. What distinguishes the virtuous person is not so much which tools they use, but when and how they choose to deploy them.

5. Conclusions About Friendship and Social Media

Many social media friendships will of course be superficial, and likewise many mediated exchanges: social media is useful for keeping in touch with casual acquaintances. But the majority of face-to-face friendships are also utility and pleasure friendships. Aristotle tells us the best kind of friendship is rare (Aristotle 1999, 1156b25). But there is no principled reason to think that it couldn't be carried out digitally, and reason to think social media offers another way to realize a difficult-to-obtain but valuable good.

We should draw several lessons about desirable features of social media and wise, prudent ways to conduct our friendships from what we have seen here. The range of conversations possible in a flexible medium like Facebook may better facilitate sharing lives than a highly constrained platform such as Instagram's photography-centric structure or Twitter's 140-character limit. If privacy is important to friendship, trustworthy and user-friendly private messaging systems may be important for friendship-conducive social media. Options for controlling how narrowly or broadly one's message is distributed, by organizing friends into "groups", as on Facebook, may prove important—or perhaps not. The serendipity of unexpected connection might be part of social media's charm. Treating posts as implicit invitations to converse, rather than broadcast announcements, may be necessary to realize certain kinds of social goods. The ability to objectively but sympathetically evaluate people based on what they say and do on social media turns out to be a valuable skill to cultivate.

Rather than fear social media as a threat to genuine friendship, we should consider how it can be used to foster an important good, by considering it in the context of the shared life characteristic of the best friendships. That is what I set out to do in the following chapters.

Bibliography

Aristotle. 1999. *Nicomachean Ethics*. 2nd ed. Translated by Terence Irwin. Indianapolis, IN: Hackett Publishing Company. https://books.google.com/books?id=-Mf5XV8q6CgC&dq.

Bank, Barbara J., and Suzanne L. Hansford. 2000. "Gender and Friendship: Why Are Men's Best Same-Sex Friendships Less Intimate and Supportive?" *Personal Relationships* 7 (1): 63–78. doi:10.1111/j.1475–6811.2000.tb00004.x.

Bloomfield, Paul. 2015. "Friendship on Facebook." In *Social Media and Living Well*, edited by Berrin Beasley and Mitchell Haney. Lexington Books.

boyd, danah. 2014. "Why Snapchat Is Valuable: It's All About Attention." *Apophenia*. Accessed March 21. www.zephoria.org/thoughts/archives/2014/03/21/snapchat-attention.html.

Briggle, Adam. 2008. "Real Friends: How the Internet Can Foster Friendship." *Ethics and Information Technology* 10 (1): 71–79. doi:10.1007/s10676–008–9160-z.

Burke, Moira, Robert Kraut, and Cameron Marlow. 2011. "Social Capital on Facebook: Differentiating Uses and Users." In 571 ACM Press. doi:10.1145/1978942.1979023.

Cocking, Dean, and Jeanette Kennett. 2000. "Friendship and Moral Danger." *The Journal of Philosophy* 97 (5): 278–96. doi:10.2307/2678396.

Cocking, Dean, and Steve Matthews. 2000. "Unreal Friends." *Ethics and Information Technology* 2 (4): 223–31. doi:10.1023/A:1011414704851.

Cohen, Sheldon, and S. Leonard Syme, eds. 1985. *Social Support and Health*. Academic Press. http://psycnet.apa.org/record/1985-97489-000.

Derlega, Valerian J., Robin J. Lewis, Scott Harrison, Barbara A. Winstead, and Robert Costanza. 1989. "Gender Differences in the Initiation and Attribution of Tactile Intimacy." *Journal of Nonverbal Behavior* 13 (2): 83–96. doi:10.1007/BF00990792.

Ellison, Nicole B., Charles Steinfield, and Cliff Lampe. 2007. "The Benefits of Facebook 'Friends:' Social Capital and College Students' Use of Online Social Network Sites." *Journal of Computer-Mediated Communication* 12 (4): 1143–68. doi:10.1111/j.1083–6101.2007.00367.x.

Fröding, Barbro, and Martin Peterson. 2012. "Why Virtual Friendship Is No Genuine Friendship." *Ethics and Information Technology* 14 (3): 201–7. doi:10.1007/s10676-011-9284-4.

Hancock, Jeffrey T. 2009. "Digital Deception: Why, When and How People Lie Online." In *Oxford Handbook of Internet Psychology*, edited by Adam N Joinson, Katelyn Y. A. McKenna, Tom Postmes, and Ulf-Dietrich Reips. Oxford University Press.

Hogluin, Jaime. 2005. "Abdullah-Bush Stroll Strikes Nerve." News. *CBS Evening News*. April 27. www.cbsnews.com/news/abdullah-bush-stroll-strikes-nerve/.

Ito, Mizuko, Heather A. Horst, Matteo Bittanti, danah boyd, Becky H. Stephenson, Patricia G. Lange, C. J. Pascoe, Laura Robinson, and Sonja Baumer. 2009. *Living and Learning with New Media: Summary of Findings from the Digital Youth Project*. MIT Press. https://books.google.com/books?id=1pxitUNb9pMC.

Lin, Liu yi, Jaime E. Sidani, Ariel Shensa, Ana Radovic, Elizabeth Miller, Jason B. Colditz, Beth L. Hoffman, Leila M. Giles, and Brian A. Primack. 2016. "Association between Social Media User and Depression among U.S. Young Adults: Research Article: Social Media and Depression." *Depression and Anxiety* 33 (4): 323–31. doi:10.1002/da.22466.

McFall, Michael T. 2012. "Real Character-Friends: Aristotelian Friendship, Living Together, and Technology." *Ethics and Information Technology* 14 (3): 221–30. doi:10.1007/s10676-012-9297-7.

Monsour, Michael. 1992. "Meanings of Intimacy in Cross- and Same-Sex Friendships." *Journal of Social and Personal Relationships* 9 (2): 277–95. doi:10.1177/0265407592092007.

Munn, Nicholas John. 2012. "The Reality of Friendship within Immersive Virtual Worlds." *Ethics and Information Technology* 14 (1): 1–10. doi:10.1007/s10676-011-9274-6.

Rainie, Lee, Barry Wellman, John B. Horrigan, and Jeffery Boase. 2006. "The Strength of Internet Ties." Pew Research Center. www.pewinternet.org/2006/01/25/the-strength-of-internet-ties/.

Søraker, Johnny Hartz. 2012. "How Shall I Compare Thee? Comparing the Prudential Value of Actual Virtual Friendship." *Ethics and Information Technology* 14 (3): 209–19. doi:10.1007/s10676–012–9294-x.

Strikwerda, Robert A., and Larry May. 1992. "Male Friendship and Intimacy." *Hypatia* 7 (3): 110–25. doi:10.1111/j.1527–2001.1992.tb00907.x.

Thomas, Laurence. 1987. "Friendship." *Synthese* 72 (2): 217–36. doi:10.1007/BF00413639.

Vallor, Shannon. 2012. "Flourishing on Facebook: Virtue Friendship & New Social Media." *Ethics and Information Technology* 14 (3): 185–99. doi:10.1007/s10676-010-9262-2.

Verduyn, Philippe, David Seungjae Lee, Jiyoung Park, Holly Shablack, Ariana Orvell, Joseph Bayer, Oscar Ybarra, John Jonides, and Ethan Kross. 2015. "Passive Facebook Usage Undermines Affective Well-Being: Experimental and Longitudinal Evidence." *Journal of Experimental Psychology: General* 144 (2): 480–88. doi:10.1037/xge0000057.

Wallace, Patricia. 1999. *The Psychology of the Internet*. New York: Cambridge University Press.

Walther, Joseph B. 2007. "Selective Self-Presentation in Computer-Mediated Communication: Hyperpersonal Dimensions of Technology, Language, and Cognition." *Computers in Human Behavior* 23 (5): 2538–57. doi:10.1016/j.chb.2006.05.002.

Wellman, Barry, Anabel Quan Haase, James Witte, and Keith Hampton. 2001. "Does the Internet Increase, Decrease, or Supplement Social Capital?" *American Behavioral Scientist* 45 (3): 436–55. doi:10.1177/00027640121957286.

Wohn, Donghee Yvette, Cliff Lampe, Rick Wash, Nicole Ellison, and Jessica Vitak. 2011. "The" S" in Social Network Games: Initiating, Maintaining, and Enhancing Relationships." In *1–10*. Kauai, HI: IEEE. doi:10.1109/HICSS.2011.400.

9 Taking Control of Conversations Through Technologically Mediated Communication

1. Introduction

Having cleared away some broad objections to technologically mediated friendship in Chapter 8, I now turn to specific concerns about particular forms of mediated interaction. I begin in this chapter by considering the ethical import of asynchronous communication to friendship, especially in light of its ability to allow friends to maintain boundaries between each other. This boundary-maintenance has struck many as contrary to the spirit of friendship. But in light of the conclusions reached in Chapter 2 about the importance of complementary differences and the conceptual possibility of valuing friends as "other selves" without presuming excessive similarity, I am skeptical of this claim. In what follows, I argue that the boundaries reinforced by asynchronous communication turn out to be highly compatible with healthy friendship.

Technologically mediated communication allows people to converse across previously insurmountable distances and time differences. One can email, text, or send a "snap"—a visual message via the social media app Snapchat—to a friend in a different time zone as easily as one down the hall, and the conversations will not appreciably differ from each other. But the conversations will differ from synchronous face-to-face conversations, and the differences between synchronous and asynchronous conversations, in particular, have been a subject of discussion for ethicists, including how this difference impacts personal relationships. In this chapter, I examine criticisms of what has come to be known in communication studies as computer-mediated communication (often abbreviated CMC) which charge that they are detrimental to friendships, and arrive at the surprising conclusion that the features previously identified as harmful actually, when properly understood, end up enhancing friendship quality and quality friendships. Before explaining why this is, however, it is helpful to get clear on what these features are.

In her book *Alone Together*, Sherry Turkle reports that 16-year-old Audrey, like many of her cohorts, much prefers texting and other asynchronous communication methods to real-time communication methods, such as the telephone.

> The phone, it's awkward. I don't see the point. Too much just a recap and sharing feelings. With a text . . . I can answer on my own time. I can respond. I can ignore it. So it really works with my mood. I'm not bound to anything, no commitment. . . . I have control over the conversation and also more control over what I say.
>
> (Turkle 2011, 190)

Audrey proceeds to explain what she prefers about her favored communication channels, coining a new word in order to better articulate her reasons. Over the telephone, she says, "there is a lot less *boundness* to the person" than when texting. By this, she seems to mean that her ability to control what she says, and when, is greatly diminished by phone as opposed to text. On a phone call, a person can demand of her things she is not comfortable giving. Callers seem to her to be imposing on her time, and even her psychological space. Turkle's diagnosis is that, for Audrey, "A call has insufficient boundaries" (2011, 190).

In this chapter, I evaluate the impact of such practices as uses of boundary-promoting communication technology. First, I investigate Friendship and the Good Life. I review reasons to think friendship is important for good human lives, as well as features that make some friendships better than others. Then, I survey previously identified Problems with Mediated Communication. Following that, I connect boundary-promoting properties of Technologically Mediated Communication and Autonomy. In that section I evaluate Marilyn Friedman's analysis of autonomy as a factor for social disruption, in order to argue that although autonomy is disruptive, it is potentially good-making for the best and most valuable friendships. I apply this analysis of Friendship Disruption and Friendship Quality to the boundary-promoting aspects of communication technologies by examining the role of Healthy Boundaries in Healthy Friendships. I conclude by noting that some of the potential shortcomings of these communication channels may be partly mitigated by technological solutions, by way of a case study of the increasingly popular app Snapchat. Its ability to mitigate some social shortcomings of technologically mediated communication may help explain the appeal of Snapchat, and suggests that we have reason to be sanguine about the long-term impacts of technologically mediated communication. People seem to intrinsically value friendship. Moreover, this valuing can motivate them to adopt technologies that promote good friendships.

2. Friendship and the Good Life

As we have seen, the concept of friendship is difficult to pin down precisely, but there are several features worth reviewing in order to properly evaluate communication technology's impact on it. First, friendship is necessarily reciprocal. It seems to be a category mistake to talk about unrequited friendship, even though individual positive dispositions toward people, such as love and affection, do not suffer from such problems. Second, friendship overlaps with other kinds of relationships. One can consider a sibling to be a friend. Likewise, one can be friends with a co-worker, or a neighbor. Furthermore, to say of some relationship that it is (also) a friendship contributes an additional positive evaluation to the relationship—to be friends with one's co-workers is to say something good about the relationship, and to say that one is not friends with a sibling is to say something bad about it.

As I do throughout this book, I draw upon Aristotle's theory of friendship, and Aristotelian virtue theory, to ground my approach. Aristotelian ethics offers some distinct advantages in evaluating the ethical impact of technology, such as its focus on the ways that even seemingly small changes to how a person lives her life can, when exercised repeatedly, impact character (Vallor 2010).

Recall Aristotle's distinction between three major kinds of friends: utility friends, pleasure friends, and virtue or character friends (Aristotle 1999; Cooper 1977). The first two kinds are, as their names suggest, primarily instrumental relationships, rooted in the logically independent goods people can provide each other, such as pleasure and utility: that is, goods that are not conceptually bound up with a particular person but could be provided by other people. My friend may be useful to me because he owns a big truck and can help me move bulky items, but anyone else with a similar truck would be similarly useful. The last form of friendship is considered by Aristotle to be the best, and is characterized by people valuing each other for their good character, their intrinsic qualities, as ends in themselves rather than means to other ends (Badhwar 1991). In what follows, I focus on communication technologies' impacts on virtue friendship, as the best and fullest form of friendship, and so the closest to our ideals about what constitutes the highest-quality personal relationships. If technologically mediated communications are harmful to virtue friendship, this is reason for concern, and if they are (or can be) beneficial for it, this should be recognized. And, in fact, many important criticisms acknowledge that instrumental friendships can be facilitated by communication technologies: the worry is that they do so only at the cost of making intrinsically valued friendships more difficult to sustain.

At first blush, it seems that friendship can be promoted by many forms of mediated communication, from instant messaging, to email, to texting, to social media ranging from Facebook to Snapchat. It enables friends to keep in touch with each other even when they are physically distant, and while some (such as Skype and FaceTime) are synchronous, many others are asynchronous. They allow people to converse even when their schedules do not align. Suppose Alice works at a bank during normal business hours, while Betty's schedule as a nurse varies wildly from day to day, with occasional overnight shifts. Alice and Betty can send each other messages whenever they have a spare moment, and each be assured that their messages will be picked up by the recipient when the other has time and energy to respond. Unlike older technologies, such as telephones, many modern communication technologies thus permit people with widely varying schedules to remain in contact with each other. Because of this, and because asynchronicity seems to introduce new challenges, in what follows I focus primarily on asynchronous communication technologies.

3. Potential Problems With Mediated Communication

Despite its apparent advantages, some aspects of asynchronous communication seem to be problematic for friendship. I focus on two major ones here. Both involve what we might think of as *moral hazards*: ways they tempt us to engage in behavior that seems convenient at the time (and possibly even conducive to friendship), but that may have deleterious consequences for both individual friendships, and one's ability to successfully enjoy rich, rewarding friendships in the long-term.

The first concern involves the way that communication technologies reward impatience, and thus seems to run the risk of making us worse friends. The worry is well articulated by Shannon Vallor (Vallor 2010; 2012). Vallor calls patience a "communicative virtue", a character trait that contributes to good, rewarding communication. Asynchronous communication technologies tend to reward impatience, thereby impeding the cultivation of patience. It is easy to exit a boring, tedious, or uncomfortable conversation—easier than it would be in a face-to-face exchange. One can click away from a chat window, swipe to another friend's page, or simply close a browser window or put away a smartphone without subjecting oneself to the kind of social pressure exerted when one's interlocutor is in the room. In fact, given the asynchronous nature of these communications and the assumption that people will drift in and out of conversations as time permits, exiting for other reasons may be all but invisible.

The immediate appeal of such communication channels is that they let us avoid difficult, uncomfortable situations, by clicking out of a chat window or navigating away from a blog post, or, as Audrey describes, by answering a text in one's own time—or not at all. Such actions may, however, harm friendships, even the best of which can require patience on occasion. And even when a particular friendship is undamaged by such actions, they give reason to be wary (Vallor 2012). They cultivate habits related to avoiding difficult conversations, which in turn impedes the development of character and so our ability to live the good life with friends.

Computer-mediated communication's tendency to undercut patience may also interfere with our ability to cultivate empathy, another communicative virtue (Vallor 2010; 2012). Empathy requires willingness to sit through difficult experiences—much like patience. This can make these technologies potentially doubly damaging, because rewarding impatience can weaken both patience and empathy. Patience is, furthermore, important for establishing that one values the whole person in a friendship, and not merely the good things they provide, Vallor argues. Impatience with others contributes to the impression that the goods one gets from one's friends are primarily instrumental. It can suggest that one will opt out when immediate tangible goods such as pleasure or utility are lacking. Patience and empathy, by contrast, show that one is concerned with—and values—the whole person, for themselves and not merely for the external goods they provide.

A related worry about instrumentality is also expressed by Turkle. "Networked, we are together", she argues,

> but so lessened are our expectations of each other that we can feel utterly alone. And there is the risk that we come to see others as objects to be accessed—and only for the parts we find useful, comforting, or amusing.
>
> (2011, 154)

This risk stems, she argues, from the ease with which many technologies allow us to cycle through our social network, in search of gratifying contact, while avoiding those who do not give us what we want, when we want it. She offers as evidence a conversation between herself and 15-year-old Ricki:

> I have a lot of people on my contact list. If one friend doesn't 'get it,' I call another". This marks a turn to a hyper-other-directedness. This young woman's contact or buddy list has become something like a list of "spare parts" for her fragile adolescent self. When she uses the expression "get it", I think she means "pick up the phone". I check with her if I have gotten this right. She says, " 'Get it,' yeah, 'pick up,' but also 'get it,' 'get me.' " Ricki counts on her friends to finish her thoughts.
>
> (Turkle 2011, 177)

This impatience with those who do not immediately "get it" can lead, Turkle fears, to the illusion that friends are good merely for scratching one's social itches.

> In a life of texting and messaging, those on that contact list can be made to appear almost on demand. You can take what you need and move on. And, if not gratified, you can try someone else.
>
> (Turkle 2011, 177)

By making it easier to "use" friends to satisfy needs and then put them aside with little or no cost when they ask for something in return, these communication channels may have a tendency to reinforce in us habits of treating friends as replaceable sources of repeatable goods, rather than irreplaceable constituents of our good, as the best form of friendship seems to require, as we saw in Chapter 1. Features that make it easy to treat friends as interchangeable, always available (in some sense), and easily dismissed when what they offer is not quite what we're looking for, seem as though they are detrimental to the best sort of friendship.

4. Technologically Mediated Communication and Autonomy

Despite these concerns, I think that asynchronous communication technologies are, on balance, beneficial to developing and maintaining high-quality friendship. In order to see why, I first discuss the ultimate source of worries about both patience and instrumentality. They are traceable to a specific kind of moral hazard: the ease with which a person can choose to both engage in and disengage from exchanges with others. They thus facilitate the enforcement of boundaries between individuals. It is one's ability to enter or exit at will that permits one to establish or maintain boundaries around others, to, as Audrey puts it, "have control over the conversation and also more control over what I say" (Turkle 2011, 190). Her appeal to control and boundaries suggests a connection between the lowered costs of conversational entry and exit, and personal autonomy.

Both Vallor's and Turkle's concerns ultimately seem to derive from these technologies' potential to enhance user *autonomy*. Autonomy, literally self-rule, can refer to a variety of things, many of which are far outside the scope of this project. In this context, what I mean is that they empower users by enhancing their communicative autonomy: their control over what, when, and how they communicate, relative to other speakers and circumstances.

Lower barriers to and lower (external) costs of boundary establishment and enforcement make it easier for individuals to choose whether—and how—to engage. People are not, as Audrey puts it, "bound to anything, no commitment". Her decisions about whether, when, and with whom to interact are made at her own discretion, on the basis of her own reasons and not subject to oversight by others (Turkle 2011, 190).

Justine Johnstone has argued that computer ethics can benefit from considering ethical implications of emerging technology in terms of its potential for empowerment (Johnstone 2007). I follow her lead here, picking up on the teenager Audrey's claim that asynchronous communication technologies give her greater control over conversations. Specifically, they lower associated costs of engagement which empower individuals to enter or exit conversations—and friendships—at will, where social pressures may otherwise keep us around. To some, this may sound like a fairly light pressure and so not a significant impediment to autonomy. But it has been documented that these social strains are not equitably distributed: women, members of racial and ethnic minority groups, younger people in the presence of older people, and other disenfranchised individuals may experience disproportionate social pressures and have correspondingly more difficulty engaging and disengaging from social interactions at will, at least in face-to-face interactions (Dovidio, Kawakami, and Gaertner 2002; Speer and Stokoe 2011). These people would thus be well-positioned to find technologically mediated communication empowering.

Specifically, these technologies seem to enhance two kinds of freedom, one negative, one positive. They provide freedom *from* coercive social pressures, by allowing people to respond at a remove from others' reactions. And they enhance freedom *to* choose when/where/with whom to engage, by providing a greater range of conversational opportunities. The autonomy involved here is largely that involved in a mere procedural conception of the phenomenon, rather than postulating more controversial substantive features of autonomy. It is also, in some sense, a relational conception of autonomy, not because it holds that people are constituted by social relations, as some have charged of robust relational accounts, but because the freedoms involved are essentially relational and interpersonal. These fairly minimal characteristics of the kind of autonomy I appeal to nonetheless have interesting implications for asynchronous communication technologies' impact on friendship.

The fact that asynchronous communication technologies empower by enhancing autonomy does not, by itself, mitigate the concerns raised earlier. Rather, it makes transparent the mechanisms by which they are created. In "Autonomy, Social Disruption, and Women", Friedman uses the painter Paul Gauguin as an example of a person who, in exercising autonomy, disrupts social bonds. Gauguin abandoned his wife and children in order to pursue his career as a painter. While he might be lauded for his commitment to artistry, he seems badly lacking on the social front. Friedman points out that he exemplifies the potential of autonomy to disrupt personal relationships—much as Vallor and Turkle worry that widespread asynchronous communication will be disruptive to friendships (Friedman 2005).

But Friedman's analysis of the mechanisms by which the exercise of autonomy disrupts relationships sheds light on the impact of asynchronous communication on friendship, suggesting both risks and benefits. She argues that reflecting on one's values can lead one to reject social connections. And exercise of autonomy can also disrupt by making her a less desirable companion to others. Autonomy thus threatens friendship by giving one power and opportunity to evaluate whether or not to preserve any given social tie, and adds an additional layer of risk by introducing the potential to change values and behavior in ways that make her an undesirable companion to her friends. This echoes the concern that asynchronous communication technology use can encourage one to become impatient and thereby a worse friend, and also that in facilitating the switch from one friend in a contact list to the next, one's social connections may be weakened or cut altogether. The autonomy enabled by technology can thus be socially disruptive. But this need not always be a bad thing.

5. Friendship Disruption and Friendship Quality

I conclude that critics are right to be concerned that asynchronous communication may disrupt and/or weaken friendships, both by making exit easier, and by allowing users to choose which messages to respond to and when. However, the ability to exit, as well as to enforce boundaries, can at the same time be a friendship enhancer in two important ways.

First, autonomy-promoting aspects of computer-mediated communication have the potential to reinforce friendships because one can choose to engage with people. As the story of Ricki, who went through her contact list until she found someone who "got" her, inadvertently illustrates, we often give up on some interactions by exchanging them for others.

Suppose one believes that the best friendships are those in which all participants choose (or would freely choose) to participate. Suppose furthermore that friendships from which a person would choose to exit, given the opportunity, tend to be suboptimal. It would then follow that even though autonomy threatens particular friendships, it does so by a sort of filtering process, in which the less-than-ideal friendships tend to be disrupted, while the best friendships tend to be reinforced. That would not guarantee that every friendship disrupted would be one that was not worth having to begin with, nor that every friendship mutually chosen would be a worthwhile one, but it would tend over time to select better ones over worse.

This is a more sanguine picture of friend-replacement than that painted by Turkle. It does not alleviate the concern that use of these technologies leads to an instrumental attitude toward friendship. But it does suggest the need for a more fine-grained analysis that avoids a false dichotomy between forced togetherness and isolating autonomy.

The original concern was that the ability to switch out connections would tend to reinforce individuals' tendency to treat friends as interchangeable sources of repeatable goods, like pleasure and utility, rather than as irreplaceable constitutive ends in themselves. I will argue that asynchronous communication technologies actually help counteract this tendency. But suppose for the moment that it is a legitimate concern. Given Aristotle's distinction between character friendship and the lesser instrumental kinds of pleasure and utility friendship, these technologies then have the potential to promote lesser forms of friendship over higher. However, it also looks as though at least some exercises of relational autonomy can reinforce social bonds by affirming particular individuals as worthy of choosing. Furthermore, users can exercise their autonomy to avoid people who make poor friends, but with whom they might otherwise engage, owing to social pressures. As we saw in Chapter 1, this is not anti-friendship but is in fact a reason grounded in our valuing of friendship itself.

When individuals reciprocally choose to interact, because each finds the interaction choice worthy, it is not obvious that we should conclude that they do so on the basis of instrumental benefit. At least some may be chosen on the basis of more friendly criteria. It is far from clear that the choices made via social media are any more likely to be made for instrumental rather than intrinsic, character-driven reasons. In fact, while there may be an initial temptation to choose on the basis of instant gratification, to the extent that this makes a person less choice worthy to others, a better grasp of long-term reciprocity and the value of listening to others may be reinforced over time, resulting in the development and sustenance of better friendships.

Furthermore, that technologies enable us to distribute social reachings-out over many friends at once may itself help counteract a potentially problematic feature of face-to-face interaction and thereby enhance intimacy. For many of us, it can be difficult to reach out to our friends for help. This reluctance often involves the feeling that one does not want to impose one's burdens on people one cares about, particularly because, assuming the other person reciprocates one's care, that person will then feel obligated to assist—and we often do not like to make our friends feel bound in this way. But when one publicly reaches out for help, the burden is distributed and its impact therefore lightened for any given individual. No one on the receiving end feels the full force of obligation while still being motivated to help, and multiple friends can more easily coordinate to assist a friend in need via channels such as those provided by broadcast social media than when communication is restricted to pairwise conversations via telephone or face-to-face. This can make it easier for us to reach out to friends when we are in need, *because* it is less burdensome on our friends. This thus promotes both vulnerability and care amongst friends.[1]

At minimum, it appears that they promote autonomy by increasing the range of choice we enjoy. We have more options when it comes to choosing which—and how many—people with whom to interact. The basis, on which the choice is made, however, is underdetermined by the evidence. Those who, like Turkle, worry that this leads to a kind of instrumentality seem to assume that (at least many of) the choices will be made on the basis of the pleasure or usefulness provided by different friends. But not all choices need be made for instrumental reasons—some could be on the basis of character, for instance, or minimizing burdens to loved ones, to name just two examples. And not every friendship is worth sticking with. If they present a moral hazard in making it easier to opt out of exchanges when they are unrewarding, they may also reinforce positive exercise of autonomy in opting out of interactions with poor companions. Thus, we have at least one reason to think that CMCs are potential good-makers for friendship, even when they are bad for particular friendships.

6. Healthy Boundaries in Healthy Friendships

A second reason to be optimistic stems from these technologies' utility as a means to enforce boundaries. The move from enforcement of boundaries via delayed response, to relationship disruption, is too hasty. Even supposing one occasionally chooses to disengage from a particular exchange and use these technologies to enforce boundaries in a particular friendship, as when someone declines to answer a text immediately, this does not necessarily pose a threat to the friendship. In fact, responding to friends when one has time and attention to do so thoughtfully, while maintaining robust interests and projects of one's own, seems to be a good thing for one's friendships. I appeal to a theory of friendships as composite social groups to explain how such groups can be strengthened by what we can think of as healthy boundaries between friends. Asynchronous communication is thus a potential good-maker for friendships and for us. Furthermore, the benefits it offers can plausibly outweigh the risks previously identified.

This is a less obvious way that low barriers to exit can enhance friendships—not merely as a sorting mechanism for screening out certain types of bad friendships (those we cannot get away from, or feel stuck with because no better option is available), but for improving the very friendships in which one occasionally exercises an option to exit.

To make sense of this idea, it will be helpful to return to the question of what makes a friendship good, and to reflect in somewhat more detail about how this impacts Aristotle's account of the best kind of friendship.

As we saw in Chapter 2, Aristotle said, repeatedly and somewhat puzzlingly, that the best friends are "other selves". One way of reading this is that Aristotle thinks friends should be very similar to each other—mirror images or twins. Cocking and Kennett offer one such interpretation (1998). This is consistent with the fact that friendship networks tend to be highly homogeneous (McPherson, Smith-Lovin, and Cook 2001). However, as a description of an ideal—what makes someone a *good* friend—it is lacking. One important good of friendship is the *different* perspectives and experiences friends contribute (Williams 1981). Too great an emphasis on similarity as a necessary quality of the best friendship rules out complementarity, the ways friendships can be enriched by differences between friends.

This problem can be avoided by adopting a different account of what it means to be another self. As was discussed in Chapter 2, we can start by thinking of friendships as composite objects in our social ontology, rather than taking friends to be our "other selves" because they are similar. Friendships, like other close-knit social groups, can be thought of as being composed of people who—like parts of a machine or organism—may differ from each other but work together as an interdependent whole. On this interpretation, friends' interactions would constitute their composing a friendship, and differences between friends, like differences between different parts, could actually enhance the functioning of the whole, allowing for a high degree of complementarity as friends contribute different strengths to their shared activities. The "parts" explanation would make friends out to be other selves by being other parts of the same whole, as we saw in Chapter 2. Aristotle's ideal, then, would turn out to prescribe interdependence and interaction without requiring similarity.

Asynchronous and spatially discontinuous exchanges enabled by technology are attractive to users precisely because they support interaction, and potentially interdependence, without requiring similarity of time or place. Thus, it can make it easier for people with different lifestyles and activities to keep in touch with each other. In this respect, the qualities that Audrey identified at the start—that she can choose when and whether to respond, according to availability and mood—look less sinister and more like features that can actually enhance very different people's ability to get along.

One reason that Audrey's comments and desire for control over her communications may look unfriendly is that there is a tendency to think that friendship involves dissolving of boundaries and merging of interests. Nancy Sherman, for instance, says of friendship that "it is a relaxing of one's own sense of boundaries and control. It is acknowledging a sense of union or merger" (Sherman 1993, 292). If this is correct, then Audrey's complaint that face-to-face conversation decreases boundaries is a complaint about a necessity for friendship. But the model I advocate here, where friends are parts of the same whole, does not imply that boundaries between friends are inimical to friendship. The reverse can be true. Internal boundaries between parts, to develop the metaphor, can lend structural integrity to the whole by enhancing each part's function on its own terms, so long as the part can continue to interact and react as appropriate.

This is consistent with how we see people, especially youth, actually using social media and other communication technologies. Many teenagers today are occupied by a variety of extracurricular activities; their busy schedules and non-overlapping but time-intensive enrichment make it difficult for them to socialize in person. Although teen users of social media prefer to meet with their friends in person, they value social media because it permits friends to keep in touch between classes, meets, rehearsals, and jobs (boyd 2014a). Although one can read this as a cost of such intensive scheduling, it is also consistent with the thought that technology benefits by permitting greater engagement with projects without sacrificing one's friendships.

While it may be tempting to idealize friendship as a selfless willingness to put one's own interests aside in order to listen to a friend, a more nuanced approach is called for. In order for friendships to be sustained and enriched by complementary companions, each person's interests must be respected and promoted. Sometimes, the complementarity that lends value to friendship may require assistance. Boundary protection can be a healthy part of mutual respect for divergent interests which themselves contribute to the value of the friendship. Vallor's point about the perils of rewarding impatience is still important. But these technologies can be beneficial to character by promoting a different kind of patience. Asynchronous communication reinforces the idea that different people are up to different things that require different schedules, and one needn't be the sort of person to put everything on hold for day-to-day conversation to be a good friend. In fact, this tendency might ultimately be bad for friendship—subservience is not necessarily a good thing in personal relationships (Hampton 1993). The ability to respond when one's schedule, mood, and attention allow one to read carefully and thoughtfully construct answers, rather than dashing off a reply as soon as a message comes in, might well help one to be a better friend.

Their ability to enforce boundaries, then, turns out to be a potentially good-making feature in several ways. It improves some friendships by facilitating personal enrichment of friends. They improve individuals' character by reinforcing respect for others' time and projects. While they are potentially disruptive to friendships, their very potential to disrupt tends to suggest problems with such friendships to begin with. All other things being equal, they are a qualified social good.

7. Snapchat

While I have defended the claim that social goods follow from the increased autonomy associated with asynchronous communication technologies, there remains one further worry. They empower individuals, but they sometimes seem to do so at the cost of equitability in friendship. Because, historically, communication technologies have made most actions besides the completed communication opaque, they provide a kind of immunity from social judgment that may present another moral hazard. This hazard can be illustrated with a classic thought experiment from the philosophical canon.

In Book II of Plato's *Republic*, Glaucon challenges Socrates to defend the value of morality with a story about the Ring of Gyges, which turns its wearers invisible. Supposedly, formerly virtuous individuals who wore the ring would behave reprehensibly once they put on the ring and could act with impunity, safe from scrutiny by others, stealing, committing adultery, and trespassing at will (Plato 1992). In many ways, asynchronous communication technologies function as communicational Rings of Gyges. When you send a message, you often have no idea when or whether it is picked up, whether the recipient skimmed and dismissed it or read carefully and sympathetically. "I didn't see the text" became the twenty-first century equivalent of "the check is in the mail". They may thereby also present the same sort of moral hazard as the Ring of Gyges.

Furthermore, once a message is sent off, it becomes, in James Moor's memorable term, "greased information"—it may be forwarded, archived, posted en masse to social media or independent websites, and otherwise passed from hand to hand at lightning speed, difficult if not impossible for the sender to control (Moor 1997).

These features may seem to contribute to the hazards identified earlier—reinforcing impatience as we flick through text messages and status updates, leading us to treat communications as commodities to be separated from their context when it is convenient to do so. But the rising popularity of a relatively new social media—Snapchat—suggests that there is market demand for technologies that help us resist such hazards, preserving what is good about technologically mediated communication while reducing some of the moral hazards of earlier technologies.

Snapchat allows users to send pictures, often overlaid with scribbled artwork, emoji, filters, or text, to each other. Once a picture is received, it can be programmed to last for only a limited time—up to 10 seconds—and then deletes itself. Users can, of course, bypass this by various means, but if Snapchat detects that someone is trying to save the image, it sends a message to this effect to the original sender, thereby removing some of the invisibility of prior communication technologies. Upon opening a "snap", its sender is automatically alerted, with further alerts provided when a recipient "replays" a snap.

It is common to suppose that the bulk of Snapchat's appeal lies in its self-deleting photos, which facilitates sexting while mitigating some of the privacy concerns historically associated with this practice. But its teen user base seems to have embraced the app for much more broad usage (boyd 2014b). Reflecting on the ways that Snapchat makes features of communication transparent that prior technologies left opaque, and the nudges these provide, it is not implausible to suppose that part of the appeal is the way it mitigates some of the moral hazards associated with earlier communication channels. As danah boyd puts it, "When someone sends you an image/video via Snapchat, they choose how long you get to view the image/video. The underlying message is simple: You've got 7 seconds. PAY ATTENTION" (boyd 2014b). The message recipient's behavior is far less opaque than with previous technologies, removing some of the Ring-of-Gyges's (Plato 1992) effect. This provides a more equitable distribution of power in the exchange. By making a message's reception transparent to the sender, some of the anxieties of sending a message off into the void are alleviated, while simultaneously taking pressure off the recipient to fire off a response simply to signal that it has been received. At the same time, it makes viewing or re-viewing the message itself a signal—perhaps reinforcing some of the subtle social pressures Vallor identifies as important to shaping communicative virtues, which have previously been neglected. In addition, it preserves boundaries in some of the positive ways discussed earlier.

This illustrates the potential for new technologies to split previously bundled concerns. When we direct our attention to the small ways that communication channels shape exchanges, they can reinforce character traits and values, while suggesting that users themselves may desire tools that maximize the quality of their friendships—and their friends. Perhaps impatience and instrumentality can be discouraged by thoughtful design features, and the rising popularity of Snapchat suggests that this may be appealing to users. I return to some of these issues in the next chapter.

8. Conclusion

I conclude that the autonomy of friends supported by asynchronous communication technologies can both enhance *and* threaten friendships. The freedom of choice that is enabled is a double-edged sword. This does not mean that its positive and negative influences cancel each other out, however. The most valuable friendships are those that are choice worthy and so reinforced by the autonomy offered by these technologies, which can also promote quality friendships by screening out those which fail on this front. While most real friendships fall somewhere between the extreme in which all interactions are freely chosen wholeheartedly, and the extreme case of a friendship where every interaction is forced, there seems to be a strong relationship between choice-worthiness of interaction and the quality of a friendship.

To the extent that they pose a moral hazard to friendships, they do so by making it easy for people to respond only to those connections they find worthwhile, and to ignore or opt out of conversations they do not find worth their time and attention. This leads people to be vulnerable in such friendships, as they can only proceed by mutual agreement and mutual engagement; either party's choosing to opt out can signal the end of a friendship. But friendships are better when they are good for each person in the friendship. Users bear the responsibility for reflecting carefully to ensure that their actions reflect their considered evaluations of which friendships, and for deciding which communications are worthy of response, and when. Vallor is correct to note that users may be tempted to give in to impatience, but this is to some degree counterbalanced by asynchronous communication's reinforcement of patience when it comes to others' projects and schedules, itself a counterbalance to concerns about instrumentality.

Furthermore, asynchronous communication technologies can strengthen extant friendships. When they enhance people's ability to advance their own well-being, friendships' quality will be enhanced, at least among those that survive the ability of participants to "opt out". This can be clarified by explaining friendships as composite social entities with friends as parts, where the interactions between parts constitute people's composing the friendship. While it is common to claim that friends "share identity", this can give the mistaken impression that friendship involves lowering or removing boundaries. The parts/whole account sketched above, by contrast, explains friends as parts of friendships, and clear boundaries or borders between parts can strengthen the integrity of the whole they compose. Both at the macro level of selecting for good friendships, and the micro level of reinforcing existing friendships, these technologies offer distinct advantages. These advantages can be reinforced with intelligent design features, as illustrated by the example of Snapchat.

Having established some value to maintaining boundaries in friendship, one also needs opportunities to support emotional interdependence. If the utility of social media were restricted to the kinds of "cool" interactions facilitated by asynchronous textual communication, as noted above, there might still be grounds for concern about the tendency of technological mediation to impede emotional engagement between friends. But as it happens, an array of features appearing in social media platforms support "warm" emotional interaction. These features are investigated in the next chapter.

Note

1 I thank Ben Fairweather for raising this point.

Bibliography

Aristotle. 1999. *Nicomachean Ethics*. 2nd ed. Translated by Terence Irwin. Indianapolis, IN: Hackett Publishing Company. https://books.google.com/books?id=-Mf5XV8q6CgC&dq.

Badhwar, Neera Kapur. 1991. "Why It Is Wrong to Be Always Guided by the Best: Consequentialism and Friendship." *Ethics* 101 (3): 483–504. doi:10.1086/293313.

boyd, danah. 2014a. *It's Complicated: The Social Lives of Networked Teens*. New Haven, CT: Yale University Press. https://books.google.com/books?id=p9u8AgAAQBAJ.

———. 2014b. "Why Snapchat Is Valuable: It's All About Attention." *Apophenia*. Accessed March 21. www.zephoria.org/thoughts/archives/2014/03/21/snapchat-attention.html.

Cocking, Dean, and Jeanette Kennett. 1998. "Friendship and the Self." *Ethics* 108 (3): 502–27. doi:10.1086/233824.

Cooper, John M. 1977. "Aristotle on the Forms of Friendship." *The Review of Metaphysics*, 619–48. doi:www.jstor.org/stable/20126987.

Dovidio, John F., Kerry Kawakami, and Samuel L. Gaertner. 2002. "Implicit and Explicit Prejudice and Interracial Interaction." *Journal of Personality and Social Psychology* 82 (1): 62–68. doi:10.1037/0022–3514.82.1.62.

Friedman, Marilyn. 2005. "Autonomy, Social Disruption, and Women." In *Feminist Theory: A Philosophical Anthology*, edited by Ann E. Cudd and Robin O. Andreasen, 339–51. Blackwell Philosophy Anthologies. Malden, MA: Blackwell Publishing. https://books.google.com/books?id=oB6EQgAACAAJ.

Hampton, Jean. 1993. "Selflessness and the Loss of Self." *Social Philosophy and Policy* 10 (1): 135–65. doi:10.1017/S0265052500004052.

Johnstone, Justine. 2007. "Technology as Empowerment: A Capability Approach to Computer Ethics." *Ethics and Information Technology* 9 (1): 73–87. doi:10.1007/s10676-006-9127-x.

McPherson, Miller, Lynn Smith-Lovin, and James M. Cook. 2001. "Birds of a Feather: Homophily in Social Networks." *Annual Review of Sociology* 27 (1): 415–44. doi:10.1146/annurev.soc.27.1.415.

Moor, James H. 1997. "Towards a Theory of Privacy in the Information Age." *ACM SIGCAS Computers and Society* 27 (3): 27–32. doi:10.1145/270858.270866.

Plato. 1992. *Republic*. Edited by C. D. C. Reeve. 2nd ed. Translated by George Maximilian Anthony Grube. Indianapolis, IN: Hackett Publishing Company. https://books.google.com/books?id=5ZjRDTmOCMoC.

Sherman, Nancy. 1993. "The Virtues of Common Pursuit." *Philosophy and Phenomenological Research* 53 (2): 277–99. doi:10.2307/2107769.

Speer, Susan. A., and Elizabeth Stokoe. 2011. *Conversation and Gender*. Cambridge: Cambridge University Press. https://books.google.com/books?id=xGgrW6sNg0UC.

Turkle, Sherry. 2011. *Alone Together: Why We Expect More from Technology and Less from Each Other*. New York, NY: Basic Books. https://books.google.com/books?id=J2ine5sIIkgC.

Vallor, Shannon. 2010. "Social Networking Technology and the Virtues." *Ethics and Information Technology* 12 (2): 157–70. doi:10.1007/s10676-009-9202-1.

———. 2012. "New Social Media and the Virtues." In *The Good Life in a Technological Age*, edited by Philip Brey, Adam Briggle, and Edward Spence, 376. Routledge Studies in Science, Technology and Society. New York, NY: Routledge. https://books.google.com/books?id=KNLngc2XBigC.

Williams, Bernard. 1981. "Persons, Character, and Morality." In *Moral Luck: Philosophical Papers 1973–1980*, edited by James Rachels, 1–19. Cambridge Paperback Library. Cambridge, UK: Cambridge University Press. https://books.google.com/books?id=wMGW2Ehldp8C.

10 What Words Can't Say

Emoji and Other Non-Verbal Elements of Technologically Mediated Communication

1. Introduction

What kind of emotional interdependence is to be found through social media? In the previous chapter, I made the case that technologically mediated communication can be useful for maintaining boundaries, an important feature of healthy relationships that are enriched by complementary differences. But one might worry that boundaries without accompanying support for emotional interconnection could be alienating. This concern is especially sharp given that in Chapter 8, I noted a concern about the emotional impoverishment of computer-mediated communication. New developments in communication technology suggest that users are embracing technologies that facilitate emotional connection. In this chapter, I argue that the kind of connection found through use of technologies differ from face-to-face interactions, but that these differences can be beneficial, especially taken in combination with the conclusions reached in the previous chapter.

The early days of computer-mediated communication were dominated by text. But recent innovations provide many options to enhance or replace text messages. These include image-sharing social media systems such as Instagram and Snapchat. It also includes to the widespread adoption of "stickers" (system-specific illustrations). Photo-sharing and photo-editing elements of text-based messenger apps are increasingly common. And Unicode Standard emoji, which operate across platforms, have risen in popularity. Even "pure" text exchanges often include data about non-verbal elements such as send time, read time, and whether the recipient(s) reviewed or saved it.

These offerings have been wildly popular. The fastest-growing social media systems in recent times have facilitated image exchange. Users report that as of 2014, upwards of 75 percent of US users of computer-mediated communications use pictographs in textual communications to friends and family (D'Aleo et al., 2015). This proportion has surely grown since then. Snapchat is one of the fastest-growing social media services at present. Its users report that they especially value it for communicating with their most intimate connections. Furthermore, they treat it as a visually enriched alternative to SMS texting (Vaterlaus et al., 2016).

This raises at least two questions. One is explanatory. What explains the appeal of these non-verbal elements of "textual" communication? The second is predictive. What sort of impact can their repeated use have on individual users and the intimate relationships where they are most used (Vaterlaus et al., 2016)? As Vallor has argued, even small-scale practices can have cumulative ethical impacts on character (Vallor 2012).

I argue that we can answer both questions by considering the nature and importance of emotional interdependence of friends. This can help us understand how these non-verbal features support such emotional interdependence. First, I argue that these communication technologies positively shape intimate relationships. Second, I argue that this same repeated use positively shapes users' self-conceptions. Given that both are benefited by these features, and that the two are collectively constitutive of friendship, I conclude that these non-verbal "textual" communications support good friendship.

I have rejected criticisms of text-based social media interactions that claim such interactions are impoverished in Chapter 8. So this might seem to be a surprising result. I agree with these critics that the shared life is an important concept for making sense of the richest and most valuable form of friendship. But they have failed to consider exactly what Aristotle says about it. As he puts it, "living together is . . . sharing of conversation and thought, not sharing the same pasture, as in the case of grazing animals" (Aristotle 1999). Text can convey thought and conversation, so it can support friends' shared lives.

But while text can support friendship, we should avoid thinking that it exhausts the range of friends' connections. A hyper-rationalist conception of thought is antithetical to the rich conception of rationality endorsed by Aristotle. Rationality, on his view, is not contrasted with emotional responses. Rather, on his view reasoning includes both logical and emotional components. The embrace of graphically-and-temporally-supplemented textual communications validates the importance of sharing both propositional content and affective elements with intimate connections. This can have salutary effects on both individuals, interactions, and their friendships. Friendships, in turn, as I argued in Part I, are fruitfully conceived of as organic entities. Like organisms, their well-being requires the thriving of individual components. It follows that this also requires that these components have healthy, mutually beneficial patterns of interaction.

2. The Impact of Non-Verbals on Intimate Relationships

First, I examine the impact of non-verbals on intimate relationships. This includes both emoji and image-intensive platforms like Snapchat. Snapchat is interesting because it is used most frequently for contact with close social connections. For example, Vaterlaus et al. found that users reported that "Snapchat is more personal" (Vaterlaus et al., 2016, 598). Piwek and Joinson concluded that Snapchat is better for bonding (intimate) than bridging (looser) social connections. They also found it is most often used to strengthen existing relationships, not form new ones (Piwek and Joinson 2016). Although my project here is not primarily an empirical one, it is worth taking account of this data as explananda. Much of it points to these technologies' promise in close-knit relationships. Ethical questions, then, include: how and in what does this promise consist? And are people correct to identify these media as beneficial to relationships? Or are they missing ethical hazards?

My answer to the first question is that the benefits of non-verbal elements of textual communication are twofold. They help with the direction of attention, and they influence affective or emotional states. To answer the second question, I consider both how these impact relationship well-being, and how they impact the well-being of individuals in such relationships. I do the latter in Section 3: Putting It All Together, and Section 4: Objections and Responses.

2.1 Attention

Attention is important for at least two reasons. Attending to our loved ones is an important part of care, which is central to intimate relationships. Additionally, in digital environments, demands on attention are many. This makes features that help direct it valuable (Broadbent and Lobet-Maris 2015). The influence of technologically mediated communication's non-verbals on attention involves three factors: temporality, shared images, and the importance of faces to the human cognitive system. I address each of these in turn.

2.1.1 Temporality

Many of the text-like communication channels on which I focus here are asynchronous. Unlike in telephone calls or videochats, users need not engage in real-time exchanges. Instead, they send and receive messages at their discretion, although sometimes within a limited time frame. (Unopened Snapchat messages, for example, may expire after 30 days.) Broadbent has found widespread preference for asynchronous over synchronous communication (Broadbent 2012). This may be because asynchronous communication allows people on different schedules, with different external demands, to remain in touch. Furthermore, they can do so without demanding attention of each other at any particular time. Nonetheless, temporal issues may arise for both sender and recipient, in a variety of ways.

First, there is the question of when messages are sent and received, and how quickly they receive responses. This issue is amplified by the fact that many current messaging systems include timestamps indicating when messages are sent and received. Quick responses can be indicative of the participants' interest in the conversation—and each other. Slower responses can show a variety of things, from disinterest to thoughtfulness. The fact that messages do not *demand* an immediate response leaves open a wide range of opportunities. This makes message timing a rich information channel for both participants. It conveys facts about how and when attention is devoted to the conversation. This can be a source of interpersonal dispute. It can make people's decisions transparent, even if not the reasons that led to them. But it does so in an environment that facilitates people's keeping in touch despite differing schedules. Thus, timing can itself be a complex factor in technologically mediated conversations.

Some communication channels also include ephemeral messaging capabilities. The most famous is Snapchat's disappearing snaps. But other platforms also include such capabilities. Google's Allo has an incognito mode allowing participants to specify how long messages in a conversation will remain visible. And Instagram has introduced a disappearing messages feature. This can have effects on the sender's comfort level with disclosing sensitive information and sentiments (more on this as follows). But here I focus on the effects of ephemeral messages on the recipient. As danah boyd has argued, this gives recipients reason to pay close attention to messages. As was discussed in the previous chapter, time-limited messages can help counteract the temptation to skim. As she memorably puts it, "When someone sends you an image/video via Snapchat, they choose how long you get to view the image/video. The underlying message is simple: You've got 7 seconds. PAY ATTENTION" (boyd 2014). She reports that teenagers often put off viewing snaps until they feel they can devote attention to absorbing their contents. They seem to prefer quality of attention over quick but superficial skimming. Many phenomena compete for our attention in technologically mediated communication (Broadbent and Lobet-Maris 2015). Our smartphones can provide a seemingly never-ending stream of notifications. Thus, there can be great value in communication channels that encourage us to pause and direct attention to a message.

Non-verbal phenomena can also influence the sender's attention. One way they do so is by making some things easier to communicate in less time. Emoji, selfies, and other quickly-conveyed images can communicate nuance, context, and emotional overtones. Such things would take far more time and characters to convey using only written language. This facilitates rich exchange that more closely approximates the possibilities of face-to-face communication. It allows us to construct digital messages with facial expressions and contextual clues provided more or less unconsciously face-to-face. Thus, it lowers the cost of conveying this emotional information.

At the same time, some features of message-construction may take *more* time and energy. Depending on how one counts, for example, there can be upwards of 2,000 emoji available to users (Davis and Edberg 2017). This requires users to consider a wide variety of possible images to send. Snapchat features hundreds of stickers. It gives users the ability to copy and paste regions of images. They can draw doodles with a range of colors. And of course one can take and retake photographs until satisfied with the final result. Said photograph can then be modified with an ever-changing array of "filters" some of which overlay the image with masks and makeup, matched to the subject's facial features. Others display graphics based on current location, speed of travel, time of day, ambient temperature, date, season, or holiday. Bitmoji provides a frequently-updated array of hundreds of images featuring one's current avatar. The Bitmoji avatar may itself be modified by changing "outfits". Users can also update the avatar's base appearance (hairstyle, facial features, physique). These and other graphic options provide means and incentive to take *more* time and devote *more* attention to tailoring messages to appeal to intimate connections. That is, these features encourage thoughtful labor in service of the relationship. And in the results, the care and time devoted to constructing a message can itself be visible to a recipient. This can count as an advantage of such phenomena, contra the widespread assumption that technological innovations are for *saving* time.

2.1.2 Image Sharing

Joint attention is an important social phenomenon. In joint attention, two or more people direct their attention at the same object and, in its highest form, are each aware of its being jointly observed (Moore and Dunham 1995). One person can invite another to engage in joint attention by pointing, following another's gaze, saying "look here", or a variety of other methods. By inviting another to attend to the same phenomenon as oneself, many shared projects are made possible. It can also be thought of as an invitation to share a focus, or a way of seeing things.

By sharing images using technologically mediated communication, senders invite recipients to engage in joint attention. The significance of this affordance is not to be underestimated. Recall Aristotle's claim that friends "share lives" by sharing "thought and conversation". Joint attention through shared images can be instrumentally valuable. It can help people to send information and coordinate activities. But furthermore, it can also (partly) constitute such sharing of lives.

In fact, the ephemerality noted earlier may help friends maintain joint attention over time. The ease with which small moments can be quickly shared and as quickly set aside can also contribute. It can encourage users to attend fully to a moment, and then move on. In discussing uses of Snapchat, Ekman argues that sharing video clips can have the effect of "affirming in a flash your existential co-orientation" despite different locations (2015, 99–100). He goes on to add that it provides "visual media intimacy with those you wish to be here now" (100). He concludes that sustained exchange of small moments that capture our attention and which we wish to share with intimate connections can provide a "surprisingly effective illusion of living continuously with each other" (U. Ekman 2015). I believe that this effect is enhanced by the fact that these images (both still and video) are ephemeral. They do not create a static photo album for friends to peruse for information *about* one's life. Instead, the sharpened attention called for by an ephemeral video invites the recipient to share in the moment *as* a moment.

Ultimately, I disagree that these exchanges provide an illusion, even an effective one. What people are doing is participating in joint attention throughout their daily lives. This seems to actually constitute their "living together" in the way that Aristotle imagines. They are both *really* looking at the same thing, albeit mediated in different ways. (And really, it is not even correct to say that each perception is mediated in a different way, given that the sender views the message before sending and thus sees it as the recipient does). A single such instance would not make for shared living, any more than spending one day in another's company. But sustained patterns of such exchanges seem to have a different status. Especially so when they are part of a larger pattern of interactions that includes a variety of content exchanges. Then, this seems like it can constitute a rich array of "shared thoughts and conversations".

2.1.3 Faces

In addition to sharing images generally, there is value specifically in sharing images of faces. Such faces can be stylized or cartoonish (as with emoji), photographed, or filmed. Many of the most common emoji are of facial expressions. And upwards of 50 percent of Snapchat messages consist of selfies (Piwek and Joinson 2016). Facial expressions can communicate a wide variety of both basic and complex emotions in compact form. They can provide context for statements (such as whether something is intended as a joke). And they can trigger sympathetic responses from the recipient. (More about this shortly.)

In fact, one criticism of computer-mediated communication has been the absence of facial expressions (Cocking and Matthews 2000). Given this, it is unsurprising that introducing images of faces has proved so popular. Unicode Standard emoji have found widespread adoption in interpersonal communication. Also, at least two aspects of smartphones enable quick and easy introduction of "selfies" to conversations. Front-facing cameras make it easy for users to photograph themselves without relying on mirrors or helpful strangers. And integrating cameras into the same device that posts to social media eliminates a step. Users no longer need to photograph with one device and then upload images to a desktop or laptop computer to share their pictures via social media. Both features make it easy and intuitive to incorporate our own facial expressions into conversations.

The importance of faces to communication in interpersonal relationships should be unsurprising. Specific regions of the brain (especially the fusiform gyrus) activate in response to faces, as opposed to other medium-sized objects (Kanwisher, McDermott, and Chun 1997). Even newborn infants preferentially track face-like visual stimuli over otherwise similar images (Johnson et al., 1991). The idea that some facial expressions are universally paired with particular emotions (P. Ekman 1970) has received widespread if not uncontested support among social scientists (Russell 1994). So it is plausible to think that facial expressions help disambiguate textual communications.

It is unsurprising that including faces increases the attention paid to messages. This is not likely the result of conscious thoughts like "this message includes a face, thus I should pay attention". Rather, it is likely the result of semi-autonomous neurological subsystems that activate in the presence of faces. We're wired to attend to them in a very basic way.

2.2 *Affective and Evaluative Influence*

In addition to impacting attention, non-verbal components of technologically mediated communication can impact users' emotional affect and evaluations. They can influence both senders and messages. Thus, they can affect the character of relationships sustained by technologically mediated communication. Here, I address three major things that can influence the character and quality of relationships: faces, humor, and, paradoxically, detachment.

2.2.1 *Faces, Again*

Besides their expressive capacities, faces, even stylized ones such as emoji, can influence our emotions and evaluative responses.

Presenting stylized faces to participants in economic games increases altruism, even when participants never meet face-to-face (Haley and Fessler 2005). The presence of eyespots increases honest behavior and reduces freeloading. When they are present, people are more likely to contribute to an honor-system collection jar for a communal coffee lounge (Bateson, Nettle, and Roberts 2006). Exchanges between participants in a two-round economic game varied depending on communication. When players communicated solely through emoji, this led to greater generosity than text-based exchanges. And both trumped results without communication.

The presence of faces in messages seems likely to change how messages are interpreted, and how recipients react to the messages. Recipients are likely to be more altruistic, honest, and generous in their responses to messages with faces than those without. They are valuable as much more than trivial entertainment. Even if each individual influence is small, selfies and emoji can dramatically change the way exchanges unfold. They allow senders to express affective content more efficiently, as noted above. But they also change the reactions of message recipients. People do not seem to explicitly reason *about* these faces. But seeing a face can change the way one weights social considerations in one's practical decision-making. This phenomenon is likely connected to the distinctive roles given to facial-processing subsystems of human brains.

And altruism, generosity, honesty, and resistance to freeloading are all *positive* features for individuals to exhibit in relationships. So the quality of relationships in which these communications are common may be improved by them.

2.2.2 Humor and Playfulness

Piwek and Joinson report that 40 percent of Snapchat users cited "desire to share funny, personal or emotional content" as a reason for using it (Piwek and Joinson 2016, 361). While this encompasses a wide variety of content possibilities, I focus here on humor. I suspect the reason users value Snapchat can be brought out by exploring its potential for humorous communication. First, the ephemeral nature of Snapchat messages may lower the stakes for users. While this has been taken to contribute to the app's potential for "sexting", it also affords other exchanges in which subjects make themselves vulnerable. It can encourage silliness among intimates. Users can share things that are salient for the shared moment but which they would prefer not to linger indefinitely.

Piwek and Joinson cite additional factors that contribute to the prevalence of playfulness. These include "quick and effortless making of selfies" and "ability to add short text comments and doodles" (Piwek and Joinson 2016, 365). Presumably the other elements of Snapchat, such as silly filters and goofy stickers, can help reinforce this tendency. They go on to say that it is "possible that the narrative, conversation-like, and intimate nature of Snapchat, with an interface that affords the easy exchange of short impressions, becomes a preferred medium to playfully socialize in a more private setting than public SNS such as Facebook". They argue that "the combination of self-destructing images with an immersive interface that restricts the scope of user interaction with the content makes Snapchat an instant narrative vehicle that is similar to verbal story exchange" (Piwek and Joinson 2016, 365)

This "easy exchange of short impressions" is also found with emoji. Kelly and Watts report that users also engage in a great deal of playful narrative exchange when using them. One subject explained,

> You just make little stories . . . you just start playing around with the emojis . . . like send a picture of a moon with a face on it, and then they would send me back like a cow, and I would send them back a turtle, and it doesn't mean anything, but it's just sort of funny . . . it eventually develops into a story. Or like a little game, where you have to like guess what they're trying to say with all the pictures.
>
> (Kelly and Watts 2015, 5)

Here, again, the wide range of expressive possibilities gives users room to explore and play. With Snapchat, the ephemerality of messages may help lead to sillier behavior. With emoji, ephemerality is not an issue. But they have an intrinsically silly and disarming appearance. Another factor that helps them encourage playful behavior may be the lack of connection between images and sender. Unlike a selfie, an array of generic facial expressions can help one to maintain some level of detachment from what is being conveyed. One may feel more comfortable "playing around" with them. There is less potential for self-consciousness or embarrassment with emoji than images associated with the self. (This theme will be picked up again in 2.2.3).

2.2.3 Detachment

This last marks an advantage technologically mediated communication can offer over face-to-face interaction. People can engage in emotionally rich exchange without some common obstacles to emotional intimacy.

Snapchat users are, according to Piwek and Joinson's survey, unlikely to be in a bad mood. Seventy-six percent of subjects surveyed reported being in a "good or very good mood" when they sent their last snap, with only 5 percent reporting that they were in a bad mood at the time (Piwek and Joinson 2016, 362). The playfulness and humor discussed above can be good for intimate relationships. Some evidence suggests that high ratios of positive to negative interactions are important for romantic relationship success (Gottman and Levenson 1999). And it would not be much of a stretch to speculate that the same could be true of platonic friendships and other intimate relationships.

But the value of these non-verbals is not exhausted by their propensity to support playful, humorous, good-feeling emotional exchange. As I hinted above, these platforms support not just *good* feelings, but room to play and explore and engage in potentially risky emotional exchanges with greater comfort and less fear of repercussions than via other communication channels.

Briggle has argued that computer-mediated friendships may enjoy an advantage over face-to-face ones. The very distance between friends that many have seen as an obstacle can encourage participants to be candid and reflective (Briggle 2008). Not having to face one's interlocutor directly can help us overcome established habits and fears. These social pressures may make it difficult to reveal feelings.

At least anecdotally, emoji have been implicated in this phenomenon. Their ability to help express difficult feelings may result partly from the distance identified above. But it seems plausible that another contributor could be the way users are presented with pre-existing array of possible emotional expressions. They invite the user to select, without judgment or ranking, whichever emoji best represents one's current emotional state.

Sternbergh describes one woman's experience using emoji to communicate with a friend about a culturally fraught and emotionally laden topic:

> Her friend Phoebe Connelly had texted her about engagement rings—a fraught subject. Connelly often addressed her engagement using emoji: the Heart, the Diamond, the Diamond Ring, the Wedding Cake, the Party Starters. (Weirdly, though, not the Bride With Veil, the most obviously wedding-related emoji, which she avoided for reasons she can't quite explain, even to herself.) "Emoji", Connelly wrote in an article for the Womanzine special emoji issue, "allow me an ironic space within the dreaded cheery sincerity of being engaged. I can emoji diamond rings; therefore, it is ok that yes, I have a diamond ring. I default to emoji, a safe argot, as a means of discussing a marriage I'm emotionally ready for, but still lack the language to describe."
>
> (Sternbergh 2014)

He also reports that emoji can help overcome long-standing habits within relationships. For example, he describes a

> woman, with a septuagenarian mother, [who] revealed to me that her mom had recently sent a text relaying regret, followed by a crying-face emoji—and that this was possibly the most straightforwardly emotional sentiment her mother had ever expressed to her.
>
> (Sternbergh 2014)

In my own experience, when I began working on this project, I described it to a colleague. This person immediately responded that his father, who has never once cried in front of him, regularly uses the crying emoji when texting. This does not appear to be an accidental feature of emoji.

Shigetaka Kurita, the creator of emoji, allegedly "looked for symbols that instantly conveyed thoughts or emotions without inspiring strong likes or dislikes in the way a picture might" when he was developing the pictographs that would constitute the first emoji (Negishi 2014).

For some of the reasons discussed above, communications via Snapchat may be subject to similar advantages. Emotionally rich technologically mediated communication may support interpersonal emotional honesty by helping users to practice what we might think of as detachment without disengagement. It helps us exercise the ability to acknowledge and communicate a wide range of feelings via playful and expressively powerful but flexible imagery. At the same time, it lowers the cost of doing so by shielding users from their interlocutors' potentially intimidating gaze and ingrained but unhelpful social norms.

3. The Impact of Non-Verbals on Self-Perception

As I noted at the outset, virtue theorists are generally committed to the idea that what we repeatedly do shapes our character. This is how both virtues and vices form. A great many of us regularly connect with our intimates via the means explored in this chapter. So the characteristics brought out or reinforced by these channels can be expected to have an impact on us. People who regularly practice paying attention to their connections, communicating their feelings, and engaging in playful exchanges may thereby start to become more attentive, better at communicating feelings, and more playful. But that is not my main focus in this section. Instead, I focus on two ways that regular participation these technologies can change the way users come to see *themselves*.

Why does this matter? The exact way that practice contributes to the development of character is an empirical matter. And the details of the sort of character developed will in any case depend on the ways that the individual uses the technology. I do not intend to endorse a simple variety of technological determinism. I am skeptical that using a particular communication technology can determinately cause changes of a particular kind to "normal" users.

Instead, I focus on two issues relating to self-perception that are enabled by the technologies in question. Both involve not just character, but how we *perceive* ourselves. Said self-perception can be built up in various ways, but ultimately contributes to the kinds of decisions we make and the ways we live our lives.

3.1 Self-Knowledge Via Emotion Acknowledgment

The first issue involves increasing knowledge of one's emotional states. When we craft messages for others about these emotions, we are their first readers. The same features that help us to communicate emotional content to others often help us as well. They give us the opportunity to clarify them to ourselves.

I start with the case of emoji. As noted above, emoji can help express socially challenging emotions. Expressing these emotions to others can help us to acknowledge them as aspects of us. This seems especially powerful when they are expressed in a form that is as visible to us as the recipient. Consider Sternbergh's case of the mother who expresses regret and inserts a crying emoji at the end of her message. She does more than communicate what she (already) knows herself to feel. She has the opportunity to see *that* she is someone who feels sad and regretful, utilizing the same content-rich and neurologically significant imagery as her daughter. In addition, the design of emoji themselves can help us come to terms with our own emotions. As noted above, Shigetaka Kurita designed them to provide users with a fairly value-neutral array of options. This enables us to express a range of emotions without thereby inspiring strong like or dislike of these feelings. This can help people introspect on which emoji in fact best express their feelings in a non-judgmental environment. The ability to identify and accept emotions without strong pro- or con-attitudes is an instance of what Gottman et al. call a beneficial kind of *meta-emotion* (feeling about feelings). "For some people", they say,

> emotions are a welcome and enriching part of their lives; they believe, in a fundamental way, that it is OK to have feelings. However, for other people, emotions are to be avoided and minimized; the world of negative emotions is seen as dangerous.
>
> (John M. Gottman, Katz, and Hooven 1996, 245)

To the extent that emoji support this way of looking at our own emotions, they may cultivate healthier meta-emotions.

Emoji may also help people overcome a tendency to think of emotions in simple binaries, as pro/con, positive or negative attitudes. The array of possible images may nudge users toward a richer accounting of varieties of experience.

So much for emoji. What about stickers and selfies? Although stickers are more complex and platform-specific, they may offer similar benefits to emoji. A wide array of graphic stickers can help us confront and convey emotions. In both cases, the images seem to be powerful because their meanings are intended to be indeterminate. This allows the user to associate them with various emotional states without bringing along much in the way of baggage.

Selfies, used as emotional punctuation in textual communication, may offer a different advantage. They are less neutral in that they are images, particularly, of the user. But their production can itself be helpful. One might think there is something artificial about conjuring up a particular emotion for a selfie. And indeed, the format is often utilized to "ham it up" and deliberately over-emote for dramatic effect. But a selfie can help the user to acknowledge the emotion they intend to convey, and reflect on how they exemplify and express this emotional state.

Users' ability to perceive of themselves as possessing a range of emotional states is, then, facilitated by use of many non-verbals in technologically mediated communication.

3.2 Perceptions of Power

All else being equal, people prefer asynchronous communication channels over synchronous ones. Much of this preference seems to be the result of power dynamics. Synchronous communication demands of another that she set aside what she's doing to respond in real time. By contrast, asynchronous communication means each party can receive and respond to messages as their activities permit. Thus, neither sender nor recipient demands nor is demanded to participate in a conversation (Broadbent 2012).

Historically, this has come with all the old disadvantages of textual communications. While the written word can in principle convey a wide range of emotions and contexts, in practice this can be difficult to achieve. Non-verbals afford users both the autonomy of asynchronous communication and the expressive powers once reserved for synchronous ones. And by giving users more time and space to refine and reflect on their emotional states, they can gain some of the advantages of distance and deliberativeness that Briggle has noted for asynchronous communication.

Lastly, many means of nonverbally conveying emotional states have historically been unconscious. Or at least, they have been difficult to control without conscious attention. Consider, for example, posture, facial expression, and tone of voice. Some are autonomic and nearly impossible for the individual to regulate, like the blush response. But emoji, stickers, and selfies put the technological equivalents of these signs under the speaker's conscious control.

Other things now under the message author's control include the angle at which one is viewed, whether one is "looked down upon" or "looked up to". When we have a choice between using a generic facial expression or a selfie, we can choose whether to reveal our race or gender to our interlocutor. We can choose how long our message is available for viewing, in platforms such as Snapchat.

Even our own images are under our control to an unprecedented degree. The genre of selfies is characterized by playful construction of self-image (Lüders, Prøitz, and Rasmussen 2010). This is enhanced by the advent of front-facing cameras, filters, image overlays, and a wide array of pictographs that can be layered over photographs. Each offers novel possibilities to both construct the self one presents to the world, and determine how and when it is viewed.

Individuals are thus empowered both socially and personally by these communication technologies.

4. Putting It All Together

Non-verbals in text communications thus have positive influence on both relationships and self-perception. Interactions with intimates are improved by their impacts on attention (of senders, recipients, and joint attention) and by encouraging expression of emotions. Attention is a valuable resource in interpersonal relationships. Expression of emotions is generally valuable. But it seems especially important in intimate relationships partly characterized by emotional interdependence. Expressed emotions make parties aware of each other's emotions, and stimulate sympathetic responses. They accomplish this last by both increased attention, and altruistic impulses promoted by features such as the use of facial imagery, which activates neural regions associated with social interactions. Thus, emotional interdependence of intimates who communicate using these tools can be enhanced. Of course, this is not in itself sufficient for promoting *high-quality* intimate relationships. People can be emotionally interdependent in unhealthy ways. This is particularly likely when responsiveness to another obscures important facts about oneself.

But coupled with their positive effects on self-perception, many of these consequences can be avoided. These technologies assist with introspecting on the quality and content of one's own emotional states. They also help us to self-author and control both our participation in relationships and how we are perceived by others. Users of the relevant communication media can then be thought of as engaging in virtuous cycles. They create reflexive feedback loops where how we see ourselves changes how we interact. This in turn changes how others interact with us. Changes to our interactions, in how others respond to us, and in how we see ourselves all follow from the process of communicating with others. Together, they enhance reflective awareness of emotions. This, in turn, is associated with healthy and thriving individuals and relationships.

5. Objections and Responses

With the main arguments now made, the reader may wonder about possible objections to and limitations of the account I have given.

For example, in this project I associate selfies with emotional interdependence and intimacy. But selfies have been empirically associated with so-called "dark triad" traits, especially narcissism (Fox and Rooney 2015; Sorokowski et al., 2015). But Piwek and Joinson point out that these results are from studies of *mass* social media, not communication between intimates (Piwek and Joinson 2016). Plausibly the behavior of individuals "on stage", so to speak, could mean very different things when they occur in more private contexts. Furthermore, the emotional interdependence results discussed here are strongest for women (especially in Piwek and Joinson's survey). Meanwhile, Sorokowski et al. found that narcissism was not correlated with selfie-posting for women, only for men. This suggests that the patterns discussed here are not correlated with narcissistic behavior. At best, the evidence shows that narcissists may be inclined to use selfies (and even then, only in one gender). It does not show that selfie use causes narcissism.

One might wonder whether emotional dishonesty is enabled by the means discussed here. For example, one might use a smiling emoji in a message when in fact one feels quite sad or angry. Or one might pose for a laughing selfie, when one actually feels quite differently. Of course, one can deceive others in person as well, adopting a fake smile or forcing a laugh. But increased control over messages and lack of independently verifiable context might enable emotional dishonesty. I grant this risk. However, I think that at least two factors can help mitigate this concern. First, the distance from social pressures afforded by technological mediation can offset the temptation to fake an emotional response. (Think of the social pressures that contribute to a tendency to offer "white lies" to keep the peace.) And second, introspectiveness is facilitated by the time and array of possible responses. This may help people to become more aware of and deliberative about their own emotional states. This should make them less likely to fake a reaction simply to satisfy social ends. It is true that the dedicated and malicious liar may find it easier to pull off a lie via Snapchat or WhatsApp than in person. But even beneficial technology does not relieve people of their responsibility to be ethical in their dealings with others. And there are other reasons to believe that lying is bad for both oneself and one's intimate relationships. These reasons should move the person tempted to lie to intimate connections in any form.

In response to the concern that emoji and selfies are trivial and superficial, I have shown that their cumulative impact is anything but. In fact, their silliness and ability to encourage playfulness are important for their benefits in intimate relationships.

One might worry that emotional content shared via technology gives the illusion of intimacy without real connection. My reply is that there is another person at the end of the connections investigated here. This is not in question. What I am doing is exploring the quality of the connection.

There may be values realizable in face-to-face exchanges that are missing from even non-verbally-enhanced technologically mediated communication. But this is not a zero-sum game. Few if any people are choosing to interact with intimates via technology rather than face-to-face. We need to be sure we're comparing the right things. We are connecting with intimates to support relationships that include in-person interactions as opportunity affords. So, the relevant comparison class is between the modes of connection that are actually available. Given these constraints, the prospects for non-verbals are promising.

One last objection runs as follows. With so many options available, the actual task of sentiment selection could be overwhelming (D'Aleo et al., 2015). First, there are technological work-arounds being developed, such as emoji-search features. If people are so overwhelmed that they do not use these resources, then the potential benefits will go unreaped. But the evidence does not suggest that people are avoiding using emoji. In fact, quite the opposite. In addition, as I argued above, there can be benefits to labor-intensive messaging resources. These technologies can help people take more time and pay more attention when crafting messages and reflecting on emotions.

6. Conclusion

In this chapter, I surveyed a variety of non-verbal elements of contemporary technologically mediated communication. I argued that their popularity is at least partly explained by their usefulness in directing attention. Furthermore, they help enhance both emotional expression and emotional responsiveness of interactions between members of intimate relationships. This leads to enhanced emotional interdependence. I went on to conclude that regular engagement with intimates in these ways fosters emotional self-awareness of users. The upshot is that these technological innovations overcome long-standing shortcomings of computer-mediated communication. At the same time, they introduce novel benefits unavailable in face-to-face interactions between intimates. They facilitate more equitable power distributions in the form of asynchronous communication. They increase deliberativeness and reflection. And they increase detachment, which allows for greater acceptance of sometimes uncomfortable emotions.

In the past several chapters, I have explored advantages and disadvantages of different forms of technologically mediated communication. In the following and final chapter, I investigate grounds for deciding amongst the various communicative options in different contexts.

Bibliography

Aristotle. 1999. *Nicomachean Ethics*. 2nd ed. Translated by Terence Irwin. Indianapolis, IN: Hackett Publishing Company. https://books.google.com/books?id=-Mf5XV8q6CgC&dq.

Bateson, Melissa, Daniel Nettle, and Gilbert Roberts. 2006. "Cues of Being Watched Enhance Cooperation in a Real-World Setting." *Biology Letters* 2 (3): 412–14. doi:10.1098/rsbl.2006.0509.

boyd, danah. 2014. "Why Snapchat Is Valuable: It's All About Attention." *Apophenia*. Accessed March 21. www.zephoria.org/thoughts/archives/2014/03/21/snapchat-attention.html.

Briggle, Adam. 2008. "Real Friends: How the Internet Can Foster Friendship." *Ethics and Information Technology* 10 (1): 71–79. doi:10.1007/s10676-008-9160-z.
Broadbent, Stefana. 2012. "Approaches to Personal Communication." In *Digital Anthropology*, edited by Heather Horst and Daniel Miller, 127–45. Berg. https://books.google.com/books?id=FVcDAAAAQBAJ.
Broadbent, Stefana, and Claire Lobet-Maris. 2015. "Towards a Grey Ecology." In *The Onlife Manifesto*, edited by Luciano Floridi, 111–24. Cham: Springer International Publishing. doi:10.1007/978-3-319-04093-6_15.
Cocking, Dean, and Steve Matthews. 2000. "Unreal Friends." *Ethics and Information Technology* 2 (4): 223–31. doi:10.1023/A:1011414704851.
D'Aleo, Francesco, Valerio Perticone, Giovanni Rizzo, and Marco Elio Tabacchi. 2015. "Can You Feel It Will You Tell Me. Encouraging Sentiment Expression on the Web." In *EAPCogSci*. http://ceur-ws.org/Vol-1419/paper0054.pdf.
Davis, Mark, and Peter Edberg. 2017. "Unicode® Technical Standard #51: Unicode Emoji." Accessed May 18. www.unicode.org/reports/tr51/#Emoji_Counts.
Ekman, Paul. 1970. "Universal Facial Expressions of Emotion." *California Mental Health Research Digest* 8 (4): 151–58. www.paulekman.com/wp-content/uploads/2013/07/Universal-Facial-Expressions-of-Emotions1.pdf.
Ekman, Ulrik. 2015. "Complexity of the Ephemeral—Snap Video Chats." *Empedocles: European Journal for the Philosophy of Communication* 5 (1): 97–101. doi:10.1386/ejpc.5.1-2.97_1.
Fox, Jesse, and Margaret C. Rooney. 2015. "The Dark Triad and Trait Self-Objectification as Predictors of Men's Use and Self-Presentation Behaviors on Social Networking Sites." *Personality and Individual Differences* 76 (April): 161–65. doi:10.1016/j.paid.2014.12.017.
Gottman, John M., Lynn Fainsilber Katz, and Carole Hooven. 1996. "Parental Meta-Emotion Philosophy and the Emotional Life of Families: Theoretical Models and Preliminary Data." *Journal of Family Psychology* 10 (3): 243. doi:10.1037/0893-3200.10.3.243.
Gottman, John Mordechai, and Robert Wayne Levenson. 1999. "What Predicts Change in Marital Interaction Over Time? A Study of Alternative Models." *Family Process* 38 (2): 143–65. doi:10.1111/j.1545-5300.1999.00143.x.
Haley, Kevin J., and Daniel M.T. Fessler. 2005. "Nobody's Watching? Subtle Cues Affect Generosity in an Anonymous Economic Game." *Evolution and Human Behavior* 26 (3): 245–56. doi:10.1016/j.evolhumbehav.2005.01.002.
Johnson, Mark H., Suzanne Dziurawiec, Hadyn Ellis, and John Morton. 1991. "'Newborns' Preferential Tracking of Face-like Stimuli and Its Subsequent Decline." *Cognition* 40 (1–2): 1–19. doi:10.1016/0010-0277(91)90045-6.
Kanwisher, Nancy, Josh McDermott, and Marvin M. Chun. 1997. "The Fusiform Face Area: A Module in Human Extrastriate Cortex Specialized for Face Perception." *Journal of Neuroscience* 17 (11): 4302–4311. doi:www.jneurosci.org/content/17/11/4302.abstract.
Kelly, Ryan, and Leon Watts. 2015. "Characterising the Inventive Appropriation of Emoji as Relationally Meaningful in Mediated Close Personal Relationships." *Experiences of Technology Appropriation: Unanticipated Users, Usage, Circumstances, and Design*. http://opus.bath.ac.uk/46780/.
Lüders, Marika, Lin Prøitz, and Terje Rasmussen. 2010. "Emerging Personal Media Genres." *New Media & Society* 12 (6): 947–63. doi:10.1177/1461444809352203.

Moore, Chris, and Philip J. Dunham, eds. 1995. *Joint Attention: Its Origins and Role in Development*. Lawrence Erlbaum Associates, Inc. https://books.google.com/books?id=Mvs_vgAACAAJ.

Negishi, Mayumi. 2014. "Meet Shigetaka Kurita, the Father of Emoji." *WSJ*. Accessed March 26. https://blogs.wsj.com/japanrealtime/2014/03/26/meet shigetaka-kurita-the-father-of-emoji/.

Piwek, Lukasz, and Adam Joinson. 2016. "'What Do They Snapchat about?' Patterns of Use in Time-Limited Instant Messaging Service." *Computers in Human Behavior* 54 (January): 358–67. doi:10.1016/j.chb.2015.08.026.

Russell, James. 1994. "Is There Universal Recognition of Emotion From Facial Expression? A Review of the Cross-Cultural Studies." *Psychological Bulletin* 115 (1): 102–41. doi:10.1037/0033-2909.115.1.102.

Sorokowski, Piotr, Agnieszka Sorokowska, Anna Oleszkiewicz, Tomasz Frackowiak, Anna Huk, and Katarzyna Pisanski. 2015. "Selfie Posting Behaviors Are Associated with Narcissism among Men." *Personality and Individual Differences* 85 (October): 123–27. doi:10.1016/j.paid.2015.05.004.

Sternbergh, Adam. 2014. "Smile, You're Speaking Emoji: The Rapid Evolution of a Wordless Tongue." Magazine. *Daily Intelligencer*. November 17. http://nymag.com/daily/intelligencer/2014/11/emojis-rapid-evolution.html.

Vallor, Shannon. 2012. "Flourishing on Facebook: Virtue Friendship & New Social Media." *Ethics and Information Technology* 14 (3): 185–99. doi:10.1007/s10676-010-9262-2.

Vaterlaus, J. Mitchell, Kathryn Barnett, Cesia Roche, and Jimmy A. Young. 2016. "'Snapchat Is More Personal': An Exploratory Study on Snapchat Behaviors and Young Adult Interpersonal Relationships." *Computers in Human Behavior* 62 (September): 594–601. doi:10.1016/j.chb.2016.04.029.

11 The Moral Import of Medium

1. Introduction

Between smartphones and web-based applications, we face an ever-increasing array of communication channels. And many of these channels are accessible via the same physical device(s). While once, one would have needed a pen, paper, and envelope (as well as a postal service and stamps) for writing a letter, a telephone for phone calls, or a telegraph office for sending telegraphs, today one can choose to videochat, email, text, Instagram, Snap, tweet, or send a WhatsApp message from the same piece of equipment—to a recipient or recipients who will be similarly equipped. The anthropologist Stefana Broadbent has observed that for contemporary users of smartphones and other ICT devices, this changes the character of decisions about how to contact any given individual. It has evolved from one based on pragmatic considerations such as cost, speed, or accessibility (does Grandma have a working telephone?) to what we can think of as a purely moral choice (Broadbent 2012, 135–37).

This choice can be fruitfully made using a virtue ethics framework, or so I will argue here. Nick Couldry puts the issue well:

> If we agree that media . . . are integral to the life conditions that humans now encounter, that is, lifeworlds of complex interconnection across large scales, then media are plausibly part of the practices that contribute to human excellence. Conducting the practice of media well—in accordance with its distinctive aims, and so that, overall, we can live well with and through media—is itself part of human excellence.
>
> (Couldry 2013, 49)

Given the importance of friendships to human flourishing, good social media use turns out to be a salient component of conducting the practice of media well. Which medium is best for the kind of conversation—and relationship—we wish to have? Amongst the smartphone-equipped, norms about appropriate (or inappropriate) channels for various kinds of conversations are arising. For instance, many think it is morally wrong, all else being equal, to break up with a romantic partner via text message. And practices such as stacking cell phones face-down on the table when dining with friends seem explicitly designed to foster face-to-face over technologically mediated exchanges.

One reaction to this range of options is to rank-order them according to their objective value for communication, or relationships. For one example, consider the intuition that we ought to prioritize face-to-face conversations over technologically mediated ones. This sort of thinking leads to injunctions such as "put down the phones and talk face to face!"

Evidence in support of this hypothesis includes both informational richness provided by shared context and non-verbal cues (Cocking and Matthews 2000), as well as character-driven considerations such as the courage to face messy human interactions unmediated, or the generosity to present oneself as willing recipient of an interlocutor's reactions, rather than shielding oneself behind a screen (Turkle 2011; 2015). This suggests that perhaps objective factors such as ability to convey non-verbal cues, availability of speakers to receive immediate and unfiltered feedback, and coincidence of context are better-making features of communication technologies, and that given the choice, we ought to prefer those means over others.

But what looks like a strength in one context can seem like a drawback in another. In a delicate and high-stakes negotiation, it might be disrespectful to push for an immediate and unfiltered response from the other party. In fact, out of consideration for the power dynamics inherent in such an exchange, it might be most thoughtful and considerate to present a job offer in the form of an email that makes explicit commitment to a particular compensation package. This method, as opposed to a face-to-face one, gives the candidate time to reflect and weigh options before responding. In other cases, information conveyed by tone and facial expression may simply be irrelevant. A simple request for a friend to bring a snack to a gathering might be more efficiently conveyed via text than coordinating a face-to-face visit. And expecting someone to copy down or remember a grocery list rather than sending it via text or a shared document might seem downright imperious. Circulating some news publicly might be hurtful, while breaking other news privately might lead to awkward questions about who's "in the know" and who isn't. Each communication channel seems to have pros and cons. And both the relative weight of strengths and drawbacks, and what counts as an advantage or disadvantage of each seems to be highly context-dependent.

This is, then, a moral decision that is not governed by the moral superiority of any particular communication channel. What, then, should we use to decide? In *Technology and the Virtues*, Shannon Vallor argues that some of the usual suspects make poor candidates for guiding decision-making involving emerging technologies (Vallor 2016). For example, one might be tempted to assume that a consequentialist approach would be best: simply choose the communication that leads to the best (independently specifiable) consequences. For example, we ought to choose email when time to deliberate and craft a response leads to the best response, or a tweet when it would be most conducive to general utility to broadcast the message. But it is very difficult to foresee all of the consequences of any given communication, let alone the cumulative effects of any given form of communication on individuals. And even when we can guess, it's not always clear whether the results are good or bad. Turkle describes several intimates who reserve arguments for text or chat programs, on the grounds that this lessens emotional reactiveness and angry reciprocation (Turkle 2015). But is this good or bad? Turkle suggests this is cowardice in the face of messy human relationships, but a case could be made that this is merely responsible management of known shortcomings—and the devil may be in the details. Either could be right, depending on individual features of character and argument content. There are just too many variables in play to reliably try to calculate net consequences.

A deontological approach would have us instead consider moral rules or principles that could guide us. But consider, as an example, the emerging norm against breaking up with someone by text. It would seem that this is a case where one ought to initiate a break-up face to face in order to show something like respect for the other person, especially by offering oneself as available to receive the other person's emotional reactions. It would seem too cold, clinical, and inhumane to break off an important relationship via text without letting the other person have their say. But this presumably is not because the content of the communication is up for debate. One is not obligated to break up in person or over the phone in order to give the other party a chance to argue one out of it. Rather, the thought seems to be something closer to the idea that, in virtue of your history together, and out of respect for the other person's feelings, to show that you care enough to hear their disappointment and grief, acknowledge it, and offer apologies when and where due. It may also serve as an important counterbalance against an all-to-common tendency to be cowardly when delivering bad or upsetting news.

But this does not seem to be a unilateral rule. A variety of factors could make this inadvisable. For example, suppose the reason one is breaking up is because the other person is starting to behave abusively. In such circumstances, it seems pragmatically inadvisable to put oneself in harm's way (especially considering the likelihood of violence when an abused partner tries to leave). Or suppose the other person isn't actively physically abusive, or threatening abuse, but treats everything as an excuse to negotiate—is this enough to give one reason to make the decision to break up unilateral in form as well as content? Suppose the other person is someone who prefers time and space to process feelings alone before discussing them with other people—should THAT factor in? That is, the fact that respect for one's partner, even at the end of a relationship, is (all else being equal) a good thing to show, is not enough to support a *rule* for action, and once all of the relevant features of context are in play it is no longer clear that one could feasibly expect a rule to govern *across* the particulars of each individual situation. At best, it is a rough rule of thumb, with a range of potential defeators, and figuring out which apply, and how much weight to give each, is a job for the skilled and wise to discern in the moment, not a formula that can be straightforwardly applied. And this is with a case where intuitions are often enough in favor of the injunction that it has arisen in popular culture! Things can only get more convoluted as we proceed to less fraught (and ritualized) day-to-day communications between those we care about. By the time we get to friendship, and its characterization by anthropologists as an "institutionalized non-institution" (Paine 1969, 514), prospects for straightforward rules or principles seem dim.

Part of the problem seems to arise from the fact that while communication technologies influence people, people's dispositions influence the ways that they use technology. Neither straightforward technological determinism nor a culture-first model will carry the day. We need to make decisions that acknowledge the potential influence of communication channels on our interactions *and* that help us create contexts in which these technologies can be used to good effect. Some of this context will be cultural, as in the enaction of norms and practices that both guide us (via rough rules of thumb) and tell us what to expect (we'd be surprised to have our doctors Snapchat us important medical information, but equally bemused to have a description of a friend's lunch shared via certified mail). Some is related to much smaller-scale norms, such as whether or not particular friends consider it rude to use a cell phone during a face-to-face interaction, which turns out to be a better predictor of relationship quality than broad cultural norms and injunctions (Hall, Baym, and Miltner 2014). But in addition to cultures (both large scale, as in "Western culture", and smaller, down to families and friend groups), individual character seems to matter. The sort of person one is must inform how one chooses. If I know I have a tendency to be cowardly when I have to deliver bad news, surely that should impact my deliberations about whether to do so in person or via text, where I do not have to confront the unhappiness that results. But if I know I tend to undervalue myself and put others first, then a decision to end a relationship that I judge to be bad for me should take that fact into account, as well. I should choose a medium that helps me to assert my boundaries. Later in this chapter we will see that some people, for example, prefer to argue over SMS or chat messenger systems as opposed to face-to-face, because they find it a better medium for making sure everyone's concerns get fair hearing without being intimidated or talked over.

When we must make sensitive judgment calls in complex social contexts, where the effects are difficult to foresee and rules of thumb are at best unreliable guides to the good, and where both cultural norms and individual character must be factored in, this suggests an approach grounded in the tradition of character ethics, Vallor argues (2016). Here, I follow her suggestion, drawing primarily on an Aristotelian model, as a robust and well-explored theoretical framework that can be fruitfully applied to complex social issues. I am not, however, interested in committing myself to a faithful reconstruction of Aristotle. Rather, I start with this tradition as a springboard.

2. Character and Communication Channels

Aristotelian character ethics starts with a theory of moral goodness as that which is appropriately related to the good life (*eudaimonia*) for human beings (Aristotle 1999). Human goodness consists, ultimately, in functioning well as the kind of thing we are: rational, social animals. Certain traits and dispositions help us to function well, both constitutively and instrumentally. Just as the trait of sharpness helps a knife to function well as a knife, certain human traits of character help us to function well as rational, social animals. The traits in which we are interested, as moral agents, are those we have some control over: not in the sense of conscious control, but as characteristics we can build up over time, through habituation. Because we can gain them, we can also lose them, and so virtues have corresponding vices, other traits into which we can be habituated that tend to impair our ability to function well as the kind of thing we are. So we aim to habituate ourselves to be courageous, not rash or cowardly; to be kind, not servile or selfish. But what constitutes the appropriate virtue as opposed to vice is not a difference in kind but in degree: courage is the appropriate amount of fear in the face of (real) danger, not a deficit of it (rashness) nor an excess (cowardice). This appropriate amount is not always the same across individuals and circumstances, but rather is "just the right amount"—between too much and too little for the person and the situation. Milo the wrestler needs more food than his scrawny assistant in order to thrive, but each has their virtuous mean with respect to diet.

Several features of the theory, as given in this brief sketch, are relevant to communication technology. First, that as rational, social animals, we want to exemplify the characteristics that help us to reason and build and maintain social bonds: these are the landmarks by which we navigate ethical decision-making. Second, that our ability to do so can be enhanced by traits that can be adjusted through time and practice—and that we want to be sure to practice well, so as not to inculcate ourselves with the wrong amounts of dispositions, such as to constitute vice instead of virtue. Third, that the virtuous amount is not a fixed target, but a context-sensitive goal that must factor in both situational and personal details. We are most of us works in progress, but we can use ideals to help us develop in healthy ways. We do not start as blank slates, but must resist some tendencies and cultivate others in order to flourish as mature moral agents.

In our case, communication technologies can shape the ways that we interact, but our pre-existing character traits also shape the ways that we use them. In turn, the ways that we use communication technologies in our reasoning and social interactions habituate us to respond in some ways rather than others, both our emotional responses and our actions, as well as our patterns of reasoning. And the character traits we want to end up with are, ultimately, those that help us to reason well and to flourish socially. Communicating with others is, of course, one of the primary ways that we both reason and interact socially, so we have every reason to choose well when it comes to communication channels: we want to go with those that both constitute good reasoning and good social interaction in the moment, as constitutive features of our living well, and those that help us develop the traits necessary to continue to flourish. As we will see, the challenge is that the influence of any given communication channel is neither good nor bad, considered unqualifiedly. Rather, each has its place, facilitating the exercise of capacities that can be part of functioning well but that also have vices of excess and deficiency. In the next section, I explore several such capacities.

3. Virtues and Vices in Communication Channels

Rather than tackle each current communication medium point by point (which seems a fairly foolish strategy, given how quickly they come and go), I consider here several dimensions along which contemporary communication channels vary, in order to assess the ethical considerations involved in locating one's choice of channel at various points on the spectra. This makes the analysis less susceptible to the passage of time, while allowing us to get a better handle on what factors contribute to the ethical choice to use, for example, Messenger, Snapchat, or Skype.

As I have argued elsewhere, asynchronous communication channels such as text and email enhance both senders' and recipients' communicative autonomy. Most obviously, they do so because an asynchronous message, such as a text, voicemail, Snap, direct message, or email, can be sent at the sender's convenience without waiting for the recipient to be available. And this is so precisely because recipients are at liberty to pick up messages whenever it suits them. This freedom to send and receive messages introduces some additional liberties. For example, when one is not bound by timing constraints, one can communicate with *anyone* who has access to the communication channel with equal ease. And when (as with many communication channels) message transmission is not geographically limited (although there are exceptions, such as Yik Yak), one can communicate with a wide range of possible conversation partners with equal ease (in theory). But one is also free to ignore messages, either altogether, or until one is prepared to deal with them. And free not just in the technical sense that one can choose when and whether to respond to them, but in addition owing to the fact that in technologically mediated communication, social pressures to respond are less, well, pressing. On a phone call or in person, one speaker can wait the other out in uncomfortable silence until someone reluctantly volunteers an answer. But via email, silence is the default state. (This is not to say that such pressures never apply, but that they are lessened via many communication channels.) In fact, in a survey of preferences for synchronous versus asynchronous communication channels, Broadbent notes that the only people to actively object to these forms of communication were those (disproportionately male) users who didn't like that they couldn't demand immediate responses! (Broadbent 2012, 139) This is one reason the ability to enforce boundaries via social media matters, as we saw in Chapter 9.

These various freedoms, to choose who, whether, and when to converse, are not intrinsically virtuous or vicious. The wise person exercises this freedom wisely, and the foolish one unwisely. But the fact that they increase or reinforce liberty is itself a factor with implications for character. Some people are overly bound up in others' opinions, servile, and overly willing to give up their interests for others, perhaps even with little sense of what they themselves value (Hampton 1993). For such people, a medium which encourages reflection and reinforces independence from others can help them to assert themselves and establish healthy boundaries. For others, perhaps those with a tendency to be a bit too self-absorbed and unsympathetic, or perhaps cowardly, the freedoms of asynchronous communication can pose a moral hazard. They can be tempted to put off difficult responses, to "ghost" on people rather than hash out disagreements, or to selectively respond only when someone meets their immediate needs rather than investing the time in a relationship that has rich rewards but takes patience to develop. The overthinker and the rash person may both find themselves at a disadvantage with asynchronous communication, the one because they can keep polishing and perfecting an email that never ultimately gets sent, the latter because they can thoughtlessly fire off missives they later regret. The question is not, then, should one use these means to communicate? Nor even, what (agent-independent) reasons dictate which form of communication is best under which circumstances? But rather, how can one wisely use these channels to cultivate healthy relationships in which one remains open to the other's perspective without losing oneself in the process? When should one use a synchronous communication channel instead? How can they play a part in cultivating and exercising agency appropriately? Here, something like Hursthouse's suggestion that one navigate other practical ethical issues by asking oneself whether one's actions would be cowardly, rash, or courageous, kind, selfish, or servile, and acting consistent with the virtues, seems to be the best tool for navigating a rich and complex problem (Hursthouse 1991).

As we might expect, synchronous communication channels offer, in many ways, the opposite benefits and costs. Talking with a person synchronously, whether by phone, videochat, or in person, is a more intrusive form of communication. It demands attention from one's conversation partner, a factor that Broadbent et al. have observed can make lower-status parties highly reluctant to reach out to higher-status individuals via these versus asynchronous communication channels. Given these facts, it might at first seem odd that emerging social norms emphasize the use of synchronous technology for serious conversation: it might seem doubly disrespectful to both communicate bad or even heartbreaking news to someone, *while* demanding that they do so at one's convenience. And furthermore, two-way synchronous communication channels do not offer a shield to the recipient, but reveal their immediate and unfiltered reactions via any information channels transmitted in that medium (sound, visuals, etc.). It might seem that not only, for example, breaking up with someone via a means that insists that they put aside a mutually acceptable time to, so to speak, be broken up with is, to put it mildly, inconsiderate. And this might seem to be compounded by the fact that the unfortunate party's grief or distress is laid bare before the bearer of the bad news.

But in synchronous communication channels, one simultaneously offers oneself as a willing recipient of their immediate and unfiltered responses. One does not try to hide from the discomfort and unhappiness the message causes. There can be courage there. And kindness, too, in making oneself a willing target of expressions of disappointment, hurt, and even accusations, being willing to accept appropriate blame by the other. At the same time, a variety of contextual features of relationship, character, and circumstance can create countervailing reasons to choose other ways to communicate. Even when one delivers weighty news, it is not always appropriate to do so in a way that either makes the other person (or people) reveal their first impressions rather than reflective conclusions. Nor is it required that in order to break bad news, one must be willing to absorb the consequences. In situations where one party is struggling with self-assertion, or where the dynamic between the two has become exploitative, discretion may be the better part of valor. It would be foolish to pretend that every rejected person's upset reactions are reasonable or appropriate to the situation. Sometimes self-respect and self-preservation are more important. And sometimes simply communicating in a way that encourages feedback merely contributes to the illusion that the conclusion is up for negotiation. Such considerations can get in the way of a clean break. Again, there is no uniform principle to be found here, but rather context-sensitive considerations better navigated by considering the relevant virtues, as well as the kind of relationship one wishes to have (or not).

Private and ephemeral communication channels such as Snapchat facilitate disclosure of sensitive information. Here, I will bend my previously stated intention to focus on frameworks rather than particular technologies in order to discuss an especially interesting case study. Snapchat, somewhat infamously, was designed in this way specifically in order to facilitate disclosure of a particular kind of sensitive information: nude photographs. (Allegedly, the ghost icon is residue from its former name, "Picaboo" (Nusca 2017).) But, curiously, Snapchat adopters did not appear to use it primarily for sexting. Usage patterns did not match the intended purpose of exchanging nude photos (Roesner, Gill, and Kohno 2014). Instead, although it has become wildly popular, it is much more closely associated with silliness than sex. In thinking about the role of selfies in social media, the benefits here become clear. When one takes a selfie and posts it to Instagram or Facebook, or some other mass-broadcast medium that displays content for extended or indefinite periods of time, one has strong incentive to choose only those pictures that show oneself in the best possible light. The very structure of communication encourages careful selection, editing, and posing. But on Snapchat, photos are shared for a few seconds with, usually, a few trusted friends. Although the technology includes the capability to share images and with one's entire contact list for 24 hours in a feature known as a "story", Snapchat users report that in practice they use it primarily for one-to-one exchanges (Piwek and Joinson 2016). (Interestingly, Instagram's version of this idea appears to be enjoying greater popularity (Ward 2017). Perhaps user expectations of a given platform matter as much as technical abilities). Selfies that are much less flattering, much lower stakes, more playful, and sillier, become easier and more comfortable to exchange. The appeal of this sort of thing can be found in Snapchat's ever-changing array of "filters" that conform to and overlay one's facial features, so one can look like a cat, a pinup girl, a golfer, or a cartoon vomiting rainbows. The ephemerality and focused reception of the app facilitate a kind of vulnerability. This structured support that enables friends to display vulnerability to each other can enrich relationships and enhance trust. Ephemeral messages are less serious, which can be a bad thing. But that very levity can enhance trust and reduce pomposity and posturing.

In contrast to ephemeral messages like Snapchat (and, less prominently, disappearing messages on Instagram and Allo), permanent instant messaging systems have been around and in popular use for longer. From SMS texting to AIM, Google suite chat programs like, GChat, Google Hangouts, and Allo, and WhatsApp Messenger, these communication channels allow instantaneous or near-instantaneous transmission of (usually text) messages to one or a few individuals. Like Snapchat, they are popular for interpersonal communications among friends, romantic partners, and families. But the permanence of message affords a very different kind of use. The fact that messages can be reviewed can have simple practical benefits: it is more convenient to text someone a short grocery list than to communicate it by phone call and place the burden on the recipient to either remember or write down the desiderata.

This reviewability offers other advantages, especially when combined with a primarily text-based channel that mediates some of the emotional feedback loops people can fall into when hashing out emotionally charged topics. Turkle describes a phenomenon she encountered, in which partners, friends, and families preferred to argue or fight via instant message rather than face-to-face. They cited two major factors that contributed to this preference: the ability to moderate emotions and remain reasonable when discussing sensitive or controversial topics, and the ability to retain a log (both for eventual reference after the fact, and with an eye to saying only things they believed would stand up to future scrutiny). "I meet families who say they like to 'talk problems out' by text or email or messaging rather than in person,' she says (2015, 29). "Some refer to this practice as 'fighting by text'. They tell me that electronic talk 'keeps the peace' because with this regime, there are no out-of-control confrontations" (2015, 29). This control is linked specifically to the medium's dampening effect on emotions. "Tempers never flare. One mother argues that when family members don't fear outbursts, they are more likely to express their feelings" (2015, 29).

A woman in her thirties lists the advantages of using texts during arguments with her partner. " 'We can get our ideas out in a cooler way. We can fight without saying things we'll regret.' And she adds another benefit: Fighting by text offers the possibility of documentation. 'If we fight by text, I have a record of what was said' " (2015, 29). Turkle charges that these people are shying away from the messy, difficult, but essential complexities of human interaction. "In all of these cases, we use technology to 'dial down' human contact, to titrate its nature and extent. People avoid face-to-face conversation but are comforted by being in touch with people—and sometimes with a lot of people—who are emotionally kept at bay. . . . *It's part of the move from conversation to mere connection*" (Turkle 2015, 29).

But I am not convinced of this. Or at least, while I grant that these channels *can* be used out of cowardice, I do not think this is universally or even mostly true. Many of us *do* struggle to keep our tempers under control, or avoid falling into reactive, unthinking patterns of response, when engaging with loved ones on loaded topics. Using tools that offer a bit of distance seems to me to sometimes help us to exercise virtues like moderation, patience, and understanding when our own tendencies would have us act differently. Even if the fully virtuous individual would not need such tools, they can help us to move from *akrasia*, or weakness of will, to continence, and to practice behaving as we wish to in our clearest moments. Furthermore, with practice, we may come to more and more naturally inhabit this role, until we no longer require such assistance. Used wisely, it would seem that such things can be important developmental tools we can use to influence our own character, and to realize the goods of mature social discourse. Furthermore, even if *we* do not need such tools, when they help our conversation partners to retain *their* composure, this can constitute a kindness and a respectful accommodation. Once again, the tool itself, although influential, is neither good nor bad. Instead, our best guide to use is to reflect on the virtues we wish to exemplify and practice, and the kinds of relationships at which to aim. And as we saw in Chapter 10, a range of tools are available for times when it is appropriate to enhance emotional engagement.

When it comes to text media used for highly public communication, Twitter is among the best known. It has played major roles in political movements such as the Iranian election protests and the Black Lives Matter movement. It has also been implicated in the rapid spread of public outcry against individuals, such as dentist Walter Palmer who killed Cecil the African lion, and Justine Sacco, the corporate communications director who tweeted a joke widely perceived to be racist just before boarding a plane to South Africa. This is not the place to delve into an in-depth exploration of the many ethical issues with Twitter. But it is worth noting, here, a couple of features that facilitate its role in mobilizing individuals en masse. A relatively low message length cap (140 characters) keeps the information content of any given message small enough that it is accessible to a wide variety of readers, from those with time to read much more to those who skim quickly in passing. Its "retweet" feature makes it quite easy for individual messages to be passed from user to user, with each retweet broadcasting to a new range of followers. And Twitter pioneered the modern use of hashtags in social media, as a way to make it easy to both tag message content as being relevant to a particular trend, and to find all messages so tagged. These can have both good and bad effects. It makes it very easy for particular incidents of racialized police brutality to gain widespread attention, by tagging text or videos with the hashtag #BlackLivesMatter. It also makes it easy for issues that incite immediate and visceral emotional reactions, regardless of their reasonableness, to spread like wildfire through its user base, as each retweet takes no more than the click of a virtual button and each new message can be fired off with a length of 140 characters or less.

Is this good or bad? Well, it is good insofar as it helps give voice to the voiceless and mobilizes people previously disenfranchized by lack of resources to help organize and collaborate. It is bad insofar as it can facilitate a kind of "mob mentality", where conversations can easily be stripped of context and "slip the leash", taking on a life of their own as they fuel public outrage. Both are real features, and both are of real concern. Knowing how and when to ethically engage with Twitter is a matter of judging correctly what kinds of messages can survive this kind of radical decontextualization, and what kinds of reactions it is good to engage with. It would be cowardly and small-minded to steer clear of political engagement, just because this involves mobilizing with others, including those with whom one may have robust disagreements. It would be equally short-sighted and self-aggrandizing to tweet away in hopes of provoking a mob reaction to those issues with which one personally disagrees, without consideration of the appropriateness of such a response. (Note that this does not mean that one is always responsible for the actions of autonomous others who choose to participate in the mob.)

Another dimension along which communication channels can vary is the extent to which they support editing and revisions prior to delivering a message. This range, from deliberateness to spontaneity, has ethical costs and benefits at both ends of the spectrum. When it comes to spontaneity, Turkle (2015) says,

> texting and email and posting let us present the self we want to be. We can edit and retouch. . . . But human relationships are rich, messy, and demanding. When we clean them up with technology, *we move from conversation to mere connection* . . .
>
> (2015, 21)

Noting a widespread preference for asynchronous rather than synchronous communication, she says, people tend to "shy away from open-ended conversations. For most purposes, and sometimes even intimate ones, they would rather send a text message than hear a voice on the phone or be opposite someone face to face" (2015, 22). The result of this preference is a restriction on possible character of conversations, she thinks:

> The anxiety about spontaneity . . . means that certain conversations tend to fall away. Most endangered: the kind in which you listen intently to another person and expect that he or she is listening to you; where a discussion can go off on a tangent and circle back; where something unexpected can be discovered about a person or an idea.
>
> (Turkle 2015, 22)

This seems to express two kinds of concerns. The first is technological. This is the worry that the temptation of a medium that allows you to retouch and edit in order to present your best self will present a significant moral hazard, a temptation to shy away from the messiness and open-endedness of human conversation. The second is character-driven. It says that many of us are *already* anxious about spontaneity because we suffer from a kind of perfectionism, one that has us constantly editing, revising, and polishing in a desire to get things "right", which in turn prevents us from responding to things in a more vulnerable but honest way.

I am uneasy about the thought that deliberation, and willingness to revise and refine, to present ourselves as and aim to be the best version of us, is always inappropriate. I am happy to grant that openness, and indeed willingness to be vulnerable, to mess up on occasion, are important characteristics for one to have if one is to enjoy the goods of a rich social life. But it is far from clear that one should thereby throw aside every opportunity to reflect, refine, and deliberate about one's communications. Nor is it clear that face-to-face conversations are always supportive of spontaneity.

In an essay defending the value of deliberateness in communication amongst friends, Adam (Briggle 2008) points out that

> offline friendships occur within complex webs of relations and social structures. These webs are freighted with demands of status, norms, expectations, and conventions that shape the nature of friendships. Friends are more or less consciously squeezed into various compromises by the structure of this overarching social ecology. It can be hard, then, to really "be myself" within any space that this web affords.
>
> (Briggle 2008, 74)

Because textual communications provide distance between oneself and one's friend, with all the social expectations and norms that we typically associate with face-to-face conversations, he argues, "handwritten or digital letters can improve sincerity" (2008, 77). But it is not just the distance that is beneficial.

> We can add to this strength another quality of written correspondence that I call deliberateness. Writing occurs at a slower pace than speaking, which fosters the attentiveness and discipline to discover deeper truths about one's nature. It also affords greater opportunities to formulate precise language to describe one's character or to articulate one's reactions to the words of a friend. This deliberateness can thus act as an anchor to submerge the friendship to greater depths, or as a brighter beam of light with which to explore aspects of our selves. In this way, friends corresponding through the written word send each other concentrated, refined, and deep indicators about who they are.
>
> (Briggle 2008, 77)

Where Turkle worried that perfectionism would interfere with our ability to explore and communicate, Briggle notes that the opposite can also be bad. Carelessness and comfort skating along the surface, sticking to the comfort of ritualized social interactions which have come to constitute well-worn grooves for us after years or decades of socialization, we can miss what we really want to say. Communication channels that force us to slow down, and give us the opportunity to revise and refine, can help us to discover more about ourselves and convey that fruitfully to others—who can in turn listen more attentively and respond in thoughtful kind.

What supports spontaneity and open-endedness can also be another way to be unreflective, to lapse unthinkingly into roles and rituals, or to hide behind a forced smile. To slow down and communicate deliberately and at a greater remove can help one to counter the forces of socialized habits, when they do not serve us, and external norms and expectations.

Surely the happy medium between perfectionism and carelessness is not a uniform calculation but a kind of thoughtful flexibility, which is itself a moving target and one where what counts as thoughtful *enough* is highly sensitive to context. Negotiating for a starting salary can be a very different affair than deciding where to go to dinner with an old friend. While both can be right (and the technology can include both affordances), what will guide wise, skillful use of the technology as opposed to short-sighted and foolish implementation is the virtue of thoughtful flexibility, both in its mature form and as a goal toward which the developing individual aims.

We can see another spectrum at work in an example Turkle previously discussed, one which also involves questions about the appropriateness of distance and deliberation versus spontaneous response. Text and text-heavy communication channels engage emotional responses less than image-heavy communication channels, audio or video communications, or of course face-to-face interactions. This can help people keep cool heads when discussing emotionally fraught topics, feel more comfortable expressing difficult emotions, and hide from messy but worthwhile interactions. To presume that emotional responses are always desirable is to undersell human nature, but to overplay the importance of rationality at the expense of emotions seems equally foolish. The trick is to use both kinds of tools as appropriate: to play up or dampen down emotional responses as appropriate. Here, again, the richness of the range afforded by technologically-enhanced communication seems a significant benefit to us.

4. How Can Virtue Ethics Help?

In the discussion above, I have emphasized that along a number of dimensions, the moral implications for various communication channels depend on both details of context, and of character. I have gestured at various virtues and vices that may be implicated, but have not yet offered a robust account of how virtue ethics can help one to make wise choices in a communicative ecosystem involving a range of morally salient features whose importance can vary in a number of ways.

Eudaimonist virtue ethics involves at least three elements that can be useful in helping us to navigate complex moral issues. First, a good theory of virtue ethics sketches out (unsurprisingly) the *virtues* we aspire to embody, by giving us a picture of the fully virtuous person, or virtuous people. We can then ask ourselves, "What would the fully virtuous moral agent do? How would they use communication technology wisely and well?" We can then use this as a goal for our own behavior: to strive to model this as closely as possible. If the fully virtuous agent is kind, courageous, and respectful, we too, should strive to use communication channels in ways that are as kind, courageous, and respectful as possible: sending a text when the information is private or irrelevant to others but useful to reference and does not require an immediate response (like a grocery list), discussing some items face-to-face when it is appropriate to offer ourselves as a sympathetic audience to the recipient's immediate and visceral reaction (when sharing tragic news about the passing of a loved one or the end of an important relationship), discussing others via email when the recipient should have time and space to consider options before responding (salary negotiations). This looks suspiciously like a list of context- but not character-specific guidelines for use of communication channels, but seems a helpful starting point for reflection.

However, this simple account of virtue theory as sketching out an ideal to which we imperfect creatures are then supposed to aspire has come under criticism on various fronts. First, as Rosalind Hursthouse notes, it may not always be helpful to imagine what the perfectly virtuous agent would do, because some circumstances in which the less-than-perfectly-virtuous find themselves are not those the fully-virtuous would encounter (Hursthouse 1991). Second, as Valerie Tiberius points out, we may have little reason to trust our own conception of what the fully virtuous agent would be like, if we ourselves are imperfect and imagining them from our own imperfect epistemic position (Tiberius 2008). Third, even if we have a clear view, it is not clear that the way forward is always to try to approximate the ideal, because this may overlook important developmental barriers. For a parallel, consider the distinction between ideal and non-ideal theory in political philosophy. In an ideal world, the concept of race might do no work in a society. But if we in our present non-ideal racialized world begin to practice "color-blindness", we may overlook important ways in which historic and contemporary practices systematically disadvantage members of racialized groups. Recognizing that the perfectly virtuous agent would already have the equanimity to discuss loaded topics face-to-face does not mean that the person struggling with controlling his temper should emulate this.

Each of these gives us reason to think that we need to consider the developing moral agent as well as the fully virtuous agent in reasoning about what to do. And this concern is shot through the various analyses of the benefits and drawbacks of different communication technologies. Asynchronous communication technology both helps enforce boundaries and enables cowardice. Revisable media both facilitate thoughtful reflection and reinforce perfectionism. Broadcast social media facilitate both political movements and mobs. We cannot use only the *good* channels, we cannot count on ourselves to always be drawn to what is best for the occasion, but neither should we reject a channel out of hand because it has vicious as well as virtuous uses. We need to think about how the virtues develop, in order to navigate around moral hazards while availing ourselves of tools that the fully virtuous might not need, in order to avoid acting viciously and practice the virtues to the degree available to us.

The first obstacle I consider is that picturing an ideal agent is not always helpful because the imperfect agent may find herself in a circumstance where the ideal agent would by definition not be. For example, the perfectly virtuous individual would never tell a lie unless it was perfectly justified. But an imperfect individual might, in a moment of weakness, tell a lie she now regrets. She cannot, then, decide how to proceed by picturing what the virtuous agent would do in her position. Hursthouse argues that in such cases, the imperfect individual may nonetheless find help from virtue ethics by considering the virtues themselves. So she can still ask herself what would be honest, thoughtful, kind, etc.: to come clean about the lie, or to leave well enough alone. And the answer to the question, "what should I do?" may depend upon the circumstances (Hursthouse 1991). This is one way that virtue ethics can help with appropriate choice of communication technology. If I know that I tend to have trouble expressing my feelings clearly and honestly in face-to-face circumstances, I may be wise to use text or email to communicate about a touchy subject, even though the perfectly virtuous agent would have no need for this. I can reach this conclusion by thinking about what would be cowardly or courageous, honest or dishonest, constructive or undermining, even if I cannot reference the virtuous agent's choices directly. If I tend to lose track of important details during conversations, a method of conversing that leaves a record may be a wise choice, even if the perfectly virtuous agent would be so attentive she needed no such aid. And, with time, by practicing the virtues in the best way available to me, I may come to more and more closely approximate the ideal agent. I may learn what to attend to in conversations and with practice come to keep a better running tally as we go. I may become more comfortable expressing myself openly and honestly, even on fraught topics. While technology can make once-difficult things easier, this is not always bad. Sometimes, we can use training wheels.

In practice this can be related to the ideal/non-ideal problem as it parallels issues in political theory. A clearer picture of what our current challenges are can help us to select appropriate tools to face them. This does not mean we should throw out information about the destination. But even with an imperfect grasp of what we're after, or a wide gap between where we are and where we want to be, attunement to clear obstacles can help us to make good decisions about what tools to use, and how to use them.

I suspect that a great deal of skepticism about the value of technologically mediated communication comes from two sorts of mistake. First is the tendency to compare technologically mediated communication with the wrong contrast cases. For example, when thinking about the value of face-to-face communication versus some technologically mediated channel, we need to be careful not to compare the best heart-to-heart in-person conversations with mediocre SMS exchanges. And in assessing what course of action is best in a given situation, it is important to keep the relevant alternatives in view.

For example, in *Reclaiming Conversation*, Turkle describes a family of a teenage girl whose parents are in the early stages of a divorce. "There is constant bickering and dinner is often the place where things come to a head. . . . When a fight blows up, as it does practically every night, Melissa's mother explodes in rage" and Melissa is drawn in. Turkle relays that

> in the chaos at dinner, with everyone yelling, cell phones come out. Her mother disappears to get support from her friends on her phone. And Melissa does her version of the same thing. She goes to her phone and to Facebook—her network.
>
> (2015, 117)

Turkle's assessment of the situation is that Melissa's mother is using her phone to avoid difficult conversations with her daughter.

> Melissa's mother is in a position to express . . . [a] commitment to her daughter, to say to her, 'This situation is bad. I'm sorry that as an adult I've put you into it. Tell me how you feel. I can't necessarily help right now, but we are in this together and I'm working to get us out of this.' Instead, she goes to her phone.
>
> (Turkle 2015, 118)

But it is not hard to imagine a situation where retreating into one's phone really is the best available alternative, given the context and character of the participants involved. In a near-ideal world, even in the midst of relationship turmoil, the mature moral agent might be able to reach out as Turkle proposes. But in the midst of an acrimonious divorce, the most this mother may be able to offer her daughter, as tempers flare, is a retreat from the worst of the fighting. And this may be no small gift.

It is well-recognized that technologies make some things easier. And in many situations, making things easier can make us worse off. For example, we may lose skills and knowledge through lack of practice, a phenomenon known as deskilling. Or we may fall prey to the moral hazard of an easy way out of an unpleasant but important task. But making things easier can also make some things more accessible, and in many cases easier to practice on a regular basis. In an ideal world we wouldn't need these activities brought within our reach, because we would already be willing and able to practice them. But in our current, imperfect state, assistance can be an important component of our development. A wide range of communicative options at our fingertips can introduce both the risks and rewards of these technologies.

The last obstacle raised above is that, given our admitted imperfections, our account of what we are aiming at may be unclear, imprecise, or simply incorrect. The problem is accentuated in two ways: first, that what counts as a good life *or* a virtuous agent is controversial—different theorists and traditions give widely differing accounts of what these could be. Second, that even if we knew what we were aiming at, we lack important information about the actual risks and rewards of emerging communication technologies. Shannon Vallor calls this "technosocial opacity" (2016, 6). So if we do not know what the effects of texting or Instagramming or Snapchatting are or will be, and we lack clear means of assessing the ultimate value of even our best guesses about these impacts, how can or should we proceed?

In response to the second challenge, Vallor argues that virtue ethics is especially well suited to the project, since it focuses on character traits it is good to have across a range of situations. I would add that by keeping an eye on these traits of character rather than relying on any particular technology or technological affordance, we are better equipped to assess the suitability of any given technology, even in the midst of imperfect information. If I know I am aiming at courage, I am already equipped to judge case by case whether any given use of a technology will help me to practice being courageous, which can help me to see and respond to emerging features of new technologies. But this requires us to solve the first puzzle. Tiberius argues that, by reframing our conception of virtue, we can focus on developing virtues that will help us navigate even without an endpoint (Tiberius 2008). And (Vallor 2016) looks for common elements among the various virtue ethics traditions, to focus on even when robust details of the good life and character are still in question. But my aim here is more humble. Whatever other features of the good life and the good person are contested, being and having a good friend seem relatively uniform features of both the accounts of the wise and the values of the many. And as noted earlier, the widest range of communication technologies are relevant amongst our closest relationships. So a good account of friendship can help us to make good decisions in this sphere, at least.

5. How Can a Good Account of Friendship Help?

I have argued at length that friendships are best thought of as complex organisms, and that the individual people who compose it are its organs. Furthermore, I have argued that the well-being of friendships is taken to be intrinsically valuable by the individuals who compose them, but that this well-being requires both that individual friends flourish, and that they be appropriately interdependent so as to constitute the friendship. Interdependence does not imply lack of boundaries, but rather coordination of parts, and the apparent blending of self-interest and interest in the other is best modeled by identifying with the whole, the friendship rather than taking friends' interests to be indistinguishable. Finally, I have argued in Chapter 3 that the well-being of individual friends requires that each be virtuous. Friendship does **not** require that individuals be similar, and in fact, insofar as differences are compatible both with coordination and interdependence, and with the well-being of individuals, they can be enriching.

The importance of both difference and inter-responsiveness can help explain the appeal of a variety of communication technologies that do not very closely resemble face-to-face interaction. To the degree that friendships are strengthened by complementary differences, as was argued in Chapter 2, we should value communication technologies that allow friends to keep in touch and share their lives with each other, without requiring them to do so while living in lockstep. In fact, friendships where friends can successfully share their lives with each other while maintaining very different lives can be especially enriching. (This is not to undersell the importance of shared interests and projects.) Ephemeral messages can build trust and create space for playfulness, as well as help friends enjoy moments *as* moments. Public messages can help different friends work from a common knowledge base when discussing mutual acquaintances. Permanent communications can give friends points of reference. Formats like email that encourage thoughtful composition can promote introspection and deliberation. Instant messaging platforms support spontaneity. And to the extent that each friend uses the full range of communication technology in ways that support their development of the virtues, the friendship is thereby strengthened, given what I have argued about the connection between good character and good friendship.

Furthermore, this account of friendship rounds out the incomplete guidance of individual virtues. Communication technologies are interpersonal. Furthermore, the widest range of communication channels tends to be used in the most intimate relationships (Haythornthwaite 2005). So decisions about which communication channel to use are very often, fundamentally, questions about how our intimate relationships are to go. It should then be unsurprising that an account of ideals of friendship can fruitfully inform our decisions about how or when to use communication channels, especially when a wide range of options are available. At any given time, I can ask myself, does this relationship require respect for differing schedules and interests? Am I failing to assert my own needs? Is it good for us to be playful? Do serious things require time and a format in which to exercise introspection? Is it most important that our decisions reflect our responsiveness to each other? Or will we be best served by considering the issue separately and reporting back with enriching but differing perspectives? Do we run the risk of blurred boundaries? Do we need to practice more sensitivity to each other? There is no one-size-fits-all answer, but these questions and an ideal of the flourishing friendship as one composed of interresponsive and flourishing individuals can be used to help us navigate difficult decisions.

This ideal is important in helping to maximize the benefits and avoid the perils of communication technologies, because it can be difficult to have one without the other. For example, Hall and Baym found that use of cell phones among close friends and intimate both "positively predicted dependence, which increased satisfaction, and positively predicted over-dependence, which decreased satisfaction" (Hall and Baym 2012, 316). In such cases, clearly, the trick is to figure out how much and what kind of connection is appropriate, and what constitutes "too much" in any given situation. One clue comes from another finding of Hall and Baym: that "entrapment, the guilt and pressure to respond to mobile phone contact, uniquely predicted dissatisfaction" in relationships (2012, 316). This suggests that an important component of the model of friendship I defend, in which the well-being of friendship requires the well-being of *both* participants, is important insofar as it provides a coherent justification for asserting oneself and establishing boundaries in the context of a relationship. In this light, many of the criticisms about the distance-making features of many communication options take on new significance. When we are always (in some sense) available to each other, maintenance of boundaries may be especially important to harmonious relationship, just as when people cohabitate. Guilt and obligation are poor bases for a thriving friendship, both empirically and conceptually.

6. Objections

Objections to the claim that the range of communication channels is best handled by appeal to contextual features considered using the framework of virtue ethics, supplemented by an account of the nature of flourishing friendship, can involve several strategies. What they have in common is an appeal to an all-things-considered weakness of some or many features of contemporary communicative options, especially in light of the full range currently available to us. This would tend to support the thesis that some modes of interaction are inherently better than others, contrary to what I have argued here. While I cannot take them all on, I engage here with several prominent ones.

The first has to do with focus. At this point, it is a truism that smartphones, laptops, and other ever-present channels for technologically mediated communication offer up a steady stream of notifications, messages, and other requests for attention. This leaves the average user in a state of perpetually negotiating multiple demands on attention. Thus, one might think, sending a WhatsApp message, even if intrinsically neutral or beneficial, is inferior to a face-to-face exchange because it will inevitably be read in a context in which the recipient is constantly engaged in task-switching and unable to give it sustained attention. My response is twofold. First, it seems a mistake to think that face-to-face interactions are distraction-free. Kids and pets needing attention, and spouses and wait staff with questions, and pots boiling over and the folks at the next table being rowdy and timers going off are all ordinary features of our interactions in various face-to-face contexts. We need to be careful not to compare the idealized face-to-face exchange with the ordinary technologically mediated one. And second, it seems a mistake even to think that all exchanges between intimates need our undivided attention for significant periods of time. In fact, quick check-ins amongst other ongoing projects can help constitute the sharing of different lives, expanding and enriching our experiences. While some conversations doubtless require our undivided attention, and it is in our interest to create minimally-distracting contexts in which to conduct them (and perhaps to create some space for exchanges to evolve in such directions, as by taking a night off away from the family to visit with a friend, or going for a hike with a loved one), this does not show that all other interactions must be inferior by extension.

The second has to do with the possibility of cognitive overload—that between competing demands on attention, and the complex calculations involved in choosing amongst so many options with so little guidance, we may simply be too overwhelmed to make good choices. This, it seems to me, is an argument in favor of utilizing virtue ethics. It is true that it would be overwhelming to calculate out all the possible consequences of each possible mode of communication each time we initiate communications. And likewise for trying to universalize one's maxim for action. But virtue ethics directs us to cultivate habits of response and traits of character that help us to become attuned to ethically salient features of context, so that the right course of action stands out for us without requiring conscious thought. By training ourselves to assess and respond in virtuous ways, we help ourselves to avoid framing such choices in ways that tend to lead to cognitive overload.

Another concern involves the ease with which people can escape from most technologically mediated communications, especially those in which reception is relatively opaque (more so for instant messaging, less so for videochat). This is the idea that character develops when we are forced to engage in situations which we'd rather escape from—building patience by having to sit through a boring conversation with an elderly relative, or a stressful exchange with an emotional friend, as was discussed in Chapter 9 and in recent literature on social media (Vallor 2010; 2012; Turkle 2011). As I have argued here and elsewhere in this volume, it seems to me a mistake to think that this implies that there is no value in exercising *more* control over our choices—this can help overcome socialized deference, institutional pressures, and the many small injustices often found in hierarchical relationships. As John Stuart Mill said in *On Liberty*, this may be a cost we are wise to bear for the sake of greater freedom (Mill 1975). But in addition, there can be benefits to making some virtues easier to practice, especially to the developing individual. Suffering may sometimes build character, but not always. Sometimes, as we saw in Chapter 9, it can be good to give people other options.

A fourth concern involves our lack of information about the ultimate effects of new technologies. What if we figure out too late that some are bad, or that having so many options is bad for us? Concerns about new technology's potential to enchant us against our better judgment, or the addictive potential of social media, seem to appeal to this sort of worry. They trade on the idea that what is initially appealing or pleasant may not be ultimately good for us, but that by the time we figure this out, we may be so entrenched that change is difficult, painful, costly, or even impossible. Evidence, for example, that using social media releases dopamine has been used to criticize it in a range of popular news articles and viral videos (Soat 2015; Weinschenk 2017; *Simon Sinek on Millennials in the Workplace* 2016). But again, it is important to make the right comparisons. Dopamine is implicated in a staggering variety of normal brain functions, especially those involved in learning and anticipation (Schultz 2007). We are social creatures, disposed to find social interactions both salient and to anticipate enjoying them. It should not be intrinsically alarming that we find technologically mediated communication to activate our cognitive systems, any more than any other kind of communication. While we cannot predict the future and have reason to proceed with our eyes wide open when using any new technology, the fact that people find technologically mediated communication interesting and rewarding does not seem to me, in and of itself, to pose any special danger. Rather, it gives us reason to consider how to use them wisely and well.

7. Conclusion

The emerging moral question faced by users of multiple communication channels is not solved by hierarchically ordering communication technologies, or by looking for context-independent rules. Rather, it is best faced by considering the interplay between technology, individual character, and interpersonal relationships. Keeping an eye on the sort of person one wishes to be, and the sort of relationships one wishes to have, can help in navigating a range of choices. Furthermore, successfully navigating such choices offers unique advantages. We are able to utilize a broader range of options to both further our own personal development and facilitate rich, rewarding relationships with loved ones, by enabling us to share lives even when we cannot share space. Although this comes with risks, the fact that people can enrich each other's lives while exploring different activities and interests is of significant value. This makes the moral choice worth facing.

Bibliography

Aristotle. 1999. *Nicomachean Ethics*. 2nd ed. Translated by Terence Irwin. Indianapolis, IN: Hackett Publishing Company. https://books.google.com/books?id=-Mf5XV8q6CgC&dq.

Briggle, Adam. 2008. "Real Friends: How the Internet Can Foster Friendship." *Ethics and Information Technology* 10 (1): 71–79. doi:10.1007/s10676–008–9160-z.

Broadbent, Stefana. 2012. "Approaches to Personal Communication." In *Digital Anthropology*, edited by Heather Horst and Daniel Miller, 127–45. Berg. https://books.google.com/books?id=FVcDAAAAQBAJ.

Cocking, Dean, and Steve Matthews. 2000. "Unreal Friends." *Ethics and Information Technology* 2 (4): 223–31. doi:10.1023/A:1011414704851.

Couldry, Nick. 2013. "Living Well with and through Media." In *Ethics of Media*, edited by Nick Couldry, Mirca Madianou, and Amit Pinchevski, 39–56. London: Palgrave Macmillan UK. doi:10.1057/9781137317513_3.

Hall, Jeffrey A., and Nancy K. Baym. 2012. "Calling and Texting (Too Much): Mobile Maintenance Expectations, (over)Dependence, Entrapment, and Friendship Satisfaction." *New Media & Society* 14 (2): 316–31. doi:10.1177/1461444811415047.

Hall, Jeffrey A., Nancy K. Baym, and Kate M. Miltner. 2014. "Put down That Phone and Talk to Me: Understanding the Roles of Mobile Phone Norm Adherence and Similarity in Relationships." *Mobile Media & Communication* 2 (2): 134–53. doi:10.1177/2050157913517684.

Hampton, Jean. 1993. "Selflessness and the Loss of Self." *Social Philosophy and Policy* 10 (1): 135–65. doi:10.1017/S0265052500004052.

Haythornthwaite, Caroline. 2005. "Social Networks and Internet Connectivity Effects." *Information, Communication & Society* 8 (2): 125–47. doi:10.1080/13691180500146185.

Hursthouse, Rosalind. 1991. "Virtue Theory and Abortion." *Philosophy & Public Affairs* 20 (3): 223–46. doi:www.jstor.org/stable/2265432.

Mill, John Stuart. 1975. *On Liberty*. Edited by David Spitz. A Norton Critical Edition. New York, NY: W. W. Norton.

Nusca, Andrew. 2017. "Snapchat: An Abridged History." *Fortune*. Accessed February 4. http://fortune.com/2017/02/04/snapchat-abridged-history/.

Paine, Robert. 1969. "In Search of Friendship: An Exploratory Analysis in 'Middle-Class' Culture." *Man* 4 (4): 505–24. doi:10.2307/2798192.

Piwek, Lukasz, and Adam Joinson. 2016. "'What Do They Snapchat about?' Patterns of Use in Time-Limited Instant Messaging Service." *Computers in Human Behavior* 54 (January): 358–67. doi:10.1016/j.chb.2015.08.026.

Roesner, Franziska, Brian T. Gill, and Tadayoshi Kohno. 2014. "Sex, Lies, or Kittens? Investigating the Use of Snapchat's Self-Destructing Messages." In *International Conference on Financial Cryptography and Data Security*, 64–76. Springer. http://link.springer.com/chapter/10.1007/978-3-662-45472-5_5.

Schultz, Wolfram. 2007. "Multiple Dopamine Functions at Different Time Courses." *Annual Review of Neuroscience* 30 (1): 259–88. doi:10.1146/annurev.neuro.28.061604.135722.

Simon Sinek on Millennials in the Workplace. 2016. 16:9 HD. Inside Quest. El Segundo, California, USA: Quest Studios. www.youtube.com/watch?v=hER0Qp6QJNU.

Soat, Molly. 2015. "Social Media Triggers a Dopamine High." *American Marketing Association*. Accessed November. www.ama.org/publications/MarketingNews/Pages/feeding-the-addiction.aspx.

Tiberius, Valerie. 2008. *The Reflective Life: Living Wisely With Our Limits*. Oxford: Oxford University Press. https://books.google.com/books?id=BpcUDAAAQBAJ.

Turkle, Sherry. 2011. *Alone Together: Why We Expect More from Technology and Less from Each Other*. New York, NY: Basic Books. https://books.google.com/books?id=J2ine5sIlkgC.

———. 2015. *Reclaiming Conversation: The Power of Talk in a Digital Age*. New York, NY: Penguin Press. https://books.google.com/books?id=OUHZCwAAQBAJ.

Vallor, Shannon. 2010. "Social Networking Technology and the Virtues." *Ethics and Information Technology* 12 (2): 157–70. doi:10.1007/s10676-009-9202-1.

———. 2012. "New Social Media and the Virtues." In *The Good Life in a Technological Age*, edited by Philip Brey, Adam Briggle, and Edward Spence, 376. Routledge Studies in Science, Technology and Society. New York, NY: Routledge. https://books.google.com/books?id=KNLngc2XBigC.

———. 2016. *Technology and the Virtues: A Philosophical Guide to a Future Worth Wanting*. New York, NY: Oxford University Press. https://books.google.com/books?id=RaCkDAAAQBAJ.

Ward, Tom. 2017. "Oh, Snap: Instagram Stories Is Killing The Competition." *Forbes*. May 3. www.forbes.com/sites/tomward/2017/05/03/oh-snap-instagram-stories-is-killing-the-competition/.

Weinschenk, Susan. 2017. "Why We're All Addicted to Texts, Twitter and Google." *Psychology Today*. Accessed September 9. www.psychologytoday.com/blog/brain-wise/201209/why-were-all-addicted-texts-twitter-and-google.

Conclusion

I conclude that social technologies, when used wisely, have the potential to support our capacity for friendship and thus enrich our lives. This is not the result of naive technological optimism, but careful consideration of our values, and a clear-minded awareness of both the potential and pitfalls of innovations in this area.

In order to innovate wisely and not merely cleverly, we face a variety of challenges. One kind of challenge arises when trying to figure out how to help someone else out, especially when that person is not currently well-equipped to make independent decisions. It can be tempting to focus on the problems we *can* solve without adequately weighing the costs of our methods of doing so, especially when the decision-makers are not the ones to bear these costs. But other problems can arise when individuals choose for themselves, not because of new temptations of emerging technologies but because of very old and well-established tendencies to prefer the flattering over the genuine, the immediately satisfying over the ultimately healthy, and other such inclinations familiar to many of us just in virtue of our humanity. At the same time, attempts to criticize others' use of social technologies, and to evaluate the pitfalls they face, can also go badly astray. Some criticisms of social media practices, for example, seem to suffer from insufficient appreciation for the often reasonable considerations that can drive users to adopt such things as asynchronous communication methods, emoji, selfies, and disappearing message systems. In order to make headway, we need to embrace a balanced approach that integrates theoretical and practical concerns, and that appropriately respects people's perspectives on their own relationships and values without taking them to be infallible.

I began this book with an investigation into some relatively abstract and technical questions about the nature of friendship. This might seem to suggest that I think of ethics of emerging technologies as top-down: first we work out the theory, and then we apply it to cases we encounter, whether in our own lives or others'. But as I hope has become clear in the latter portions of the book, I see this as very much a bidirectional process. It is often helpful to approach practical problems with some ethical theory in hand, but in exploring the details of a problem we often encounter results that lead us to expand, clarify, or revise our starting theory.

By looking at how social robots are being implemented and used, and by examining social media users' practices, we can come to better understand friendship itself. Just as thought experiments can help us to clarify and organize our ethical intuitions, emerging technologies offer a unique opportunity to see how real-life "thought experiments" play out. We don't have to suppose, "what if someone could consistently make you feel like you had a friend, without caring for you, and yet you could be assured that they were not doing so in order to exploit you?" As we saw in Chapter 5, such things already exist and are being deployed in nursing homes. Or, "what if you could share *just* your thoughts and experiences with your friends, without needing to spend time with them—would this let us really share our lives?" In Chapter 8, we encountered a variety of social media platforms that raise just such a question for millions of us. While ethical theory can help us to engage in ethical practices involving social technologies, emerging social technologies can help us to refine our theories and understanding of ethical phenomena as old and ubiquitous as friendship.

Considering these questions in the abstract would run the risk of any thought experiment: it might be uninformative because we would be unable to effectively imagine something so strange and different, or because we might illegitimately smuggle in details of context that would make a difference, without realizing it. But when these technologies are being developed and deployed in particular social contexts, we can investigate with a more open mind, and use our findings to both clarify theory and structure practice. For example, in Part II, a detailed investigation into social robotics uncovered a rich variety of uses for the appearance of friendship. Some, such as those that help us to exercise social skills, especially in contexts where real social connections are difficult to ensure for a variety of reasons, turn out to offer benefits in the social realm. Others, such as those that have us treat social interests as itches to be scratched by any means necessary, end up devaluing friendship and obscuring important information about the relationship between reality and appearances.

Because some of the beneficial uses of social robotics are therapeutic, while others involve individual consumers, we also ended up investigating legitimate grounds for paternalism and limitations to its utility. Geriatric patients may mistake their relationships with friendly-looking robots for the sorts of social connections they have historically associated with appearances of friendship, while children may not learn the distinction between appearance and reality unless they are introduced correctly. And even relatively competent adults may overlook the difference when the cost of making human connections is high enough. This ended up invoking a range of considerations about our own and others' abilities to judge matters of friendship. I concluded that these were not wholly novel problems, but can be located on a continuum that includes more familiar choices about how we make decisions about friendship. But by exploring one end of the continuum, we learn something about the extremes, the similarities, and points of difference, that can inform our understanding of friendship overall.

The relationship between social media and friendship is more familiar to many of us in our daily lives. But despite the ubiquity of social media, its effects on our relationships are quite tricky to sort out. Among other things, this stems from the fact that social media users are often at least as innovative as its designers. Their practices, then, deserve attention in their own right, not merely to fuel speculation about ways that things might go wrong, but to get at the problems they are trying to solve, ones that we may have overlooked from the armchair. For example, the widespread popularity of asynchronous communication channels highlights the power imbalance issues that infect many of our interpersonal relationships, and show us what respect for others' lives and projects can look like in the context of efforts to maintain connections. And the surprising popularity of Snapchat and emoji emphasize the importance of the ephemeral, of playfulness and a lighthearted acceptance of emotions, features that can tend to be overlooked in the somewhat dour and serious-minded world of academic study of ethics and social relations.

Theorists can miss these benefits when they fear change, mistaking differences in interaction patterns for deterioration of social fabric. The fact that friendship is an established object of social value, and that longstanding practices have been used to support it, does not mean that any deviation from tradition must be a loss. Some ill-supported criticisms, including those based on misunderstandings of neurobiology—like superficial appeals to the dangers of dopamine—have enjoyed disproportionate attention in the popular press. Our tendency to exercise cleverness in place of wisdom may be partly to blame here. It can be as tempting to think we've spotted the flaws in someone else's way of life (and a ready solution for their problems) as to think that we've invented a clever new gadget that will solve social ills.

This does not mean that users are always right. It does not even mean that the most popular option among users is the right one. I take seriously Aristotle's caution that, when looking for answers to questions of value, "the many do not give the same answer as the wise" (Aristotle 1999, 3, 1095a20). Nor does it mean that deliberate design features have no impact on the quality of connection available to them. And we should heed lessons from history about the sneaky, unexpected ways that technologies can change our lives before we fully notice what's happening, in ways that are not always for the best. The automobile gave rise to the suburbs and all of the attendant social, psychological, economic, and environmental impacts that followed. And while many individuals live farther from work to save on housing, trading a longer commute for more affordable homes, this does not seem to be making them happier, a fact that puzzles economists (Stutzer and Frey 2008).

I do not assume that the invisible hand of the market will always lead us to the best results, individually or collectively. But at the same time, I think that by reflecting on people's practices, we can often get a better sense of what we value, and what kinds of tradeoffs are in play, more so than we can imagine in even the most careful thought experiment. In this case, it would seem that respecting others' time and projects, maintaining connections with those leading different sorts of lives not always conducive to face-to-face interaction, and a kind of lighthearted emotional intimacy emerge as important ingredients for good friendships, and ones that can be supported by thoughtful use of social technologies.

This is unsurprising once we observe that we are both technologically innovative and deeply social creatures. It is both unsurprising that we would look for ways to develop and use technologies as efficient tools to promote what we care about, and that we find ways to enjoy being social via creative technological means. Social technologies are expressions of two important facets of our natures. Neither of these logically requires the other. But it is to be expected that we will enjoy integrating them where we can, finding creative new ways to express affection, exercise empathetic responses, care for each other, and share things we care about together.

This naturalness still leaves us vulnerable to very human flaws. Of particular relevance here is a tendency to pursue clever solutions even when they are not ultimately good for us. One thing we have seen repeatedly in this volume is the danger of focusing too narrowly on a part of sociality without taking the bigger picture into consideration. Whether focusing on a particular cost, such as the temptation to escape from taxing conversations, or a specific benefit, such as the health gains associated with alleviating a subjective feeling of loneliness, smart people can spin out elaborate systems that take these concerns as starting premises but overlook the importance of autonomy in healthy relationships, or of genuineness in social connections. Wisdom, by contrast, involves both identifying particular details and appreciating how they add up to a more complex picture of good relationships.

It is this concern for keeping the bigger picture in view that has driven my interest in an ontology of friendship. An account of how friends relate to each other, and how their interests overlap, diverge, and interrelate, can help us to think carefully about a slippery subject. In turn, by thinking about what constitutes excellence in friendship, we can better understand what traits we want to cultivate in ourselves and secure in our social connections, in order to enjoy something of great value. Cultivating these virtuous dispositions will help us to exercise good judgment in our uses of technology. It is this last point that makes me, ultimately, optimistic about social technologies. Because these are areas of living at which we can excel, and where excellence in these areas seems connected to excelling at living as the kind of creatures we are, we have good reason to reflect on what it is to innovate wisely and well when it comes to social technologies, and good reason to put these reflections into practice, thereby helping us to live well.

Bibliography

Aristotle. 1999. *Nicomachean Ethics*. 2nd ed. Translated by Terence Irwin. Indianapolis, IN: Hackett Publishing Company. https://books.google.com/books?id=-Mf5XV8q6CgC&dq.

Stutzer, Alois, and Bruno S. Frey. 2008. "Stress That Doesn't Pay: The Commuting Paradox." *Scandinavian Journal of Economics* 110 (2): 339–66. doi:10.1111/j.1467–9442.2008.00542.x.

Index

For Product Safety Concerns and Information please contact our EU
representative GPSR@taylorandfrancis.com
Taylor & Francis Verlag GmbH, Kaufingerstraße 24, 80331 München, Germany

www.ingramcontent.com/pod-product-compliance
Ingram Content Group UK Ltd.
Pitfield, Milton Keynes, MK11 3LW, UK
UKHW022118190926
475702UK00013B/1509
* 9 7 8 0 3 6 7 8 8 9 4 3 2 *